THING EXPLAINER

COMPLICATED STUFF IN SIMPLE WORDS

THING EXPLAINER

COMPLICATED STUFF IN SIMPLE WORDS

RANDALL MUNROE

JOHN MURRAY

First published in Great Britain in 2015 by John Murray (Publishers)

An Hachette UK Company

4

A CIP catalogue record for this title is available from the British Library

ISBN 978-1-47362-091-9

Book design by Christina Gleason

Printed in China

John Murray policy is to use papers that are natural, renewable and recyclable products and made from wood grown in sustainable forests. The logging and manufacturing processes are expected to conform to the environmental regulations of the country of origin.

John Murray (Publishers)
Carmelite House
50 Victoria Embankment
London EC4Y 0DZ

www.johnmurray.co.uk

THINGS IN THIS BOOK BY PAGE

PAGE BEFORE THE BOOK STARTS

Hi!

This is a book of pictures and simple words. Each page explains how something important or interesting works, using only the ten hundred words in our language that people use the most. This page is here to say hello and explain why the book is like this.

I've spent a lot of my life worried that people will think I don't know enough. Sometimes, that worry has made me use big words when I don't need to.

One thing that I've sometimes used big words for is the shape of the world. The world is round, but it's not *exactly* round. Because of how it spins, it's a little wider around the middle. If you're building a space boat that's going to fly around the world, you have to be clear about what shape the world is, and there are some big words that you can use instead of "round." But most of the time, it doesn't matter exactly what the shape is, so people just say "round."

When I was in school, I learned about space boats and learned to use lots of big words for things like the shape of the world. Sometimes I would use those big words because they were different from the small words in an important way. But a lot of the time, I was really just worried that if I used the small words, someone might think I didn't know the big ones.

I liked writing this book because it made me let go of my fear of sounding stupid. After all—when you're saying things like "space boats" and "water pushers," *everything* sounds stupid. Using simple words let me stop worrying so much. I could just have fun making up new names for things and trying to explain cool ideas in new ways.

Some people say that there's no reason to learn big words in the first place—all that matters is knowing what things *do*, not what they're *called*. I don't think that's always true. To really learn about things, you need help from other people, and if you want to understand those people, you need to know what they mean by the words they use. You also need to know what things are called so you can ask questions about them.

But there are lots of other books that explain what things are called. This book explains what they do.

Okay, I'm done talking about the book now. Turn the page to learn about space!

SHARED SPACE HOUSE

This building flies through space just above the air. People from different countries built it and fly up to visit it in space boats.

Because the house is falling around the Earth, things inside it hang in the air instead of dropping to the floor. Inside the house, normal things like water act very strange, and you can fly around by kicking off the walls. Everyone says it's a lot of fun.

The people in the house spend their time working, playing, and taking pictures of Earth. They do work for people on the ground, helping to learn how things like flowers and machines work in space. Most of the time, there are six people in the house, with each person staying for half a year.

A big reason we built the space house was so we could learn to keep people alive and strong in space for months or years without getting sick. We'll need to be good at that if we ever want to travel to other worlds.

To build the space house, we took each piece up in a boat, pushed it until it went really fast, then caught up to the house, and stuck the part to the house.

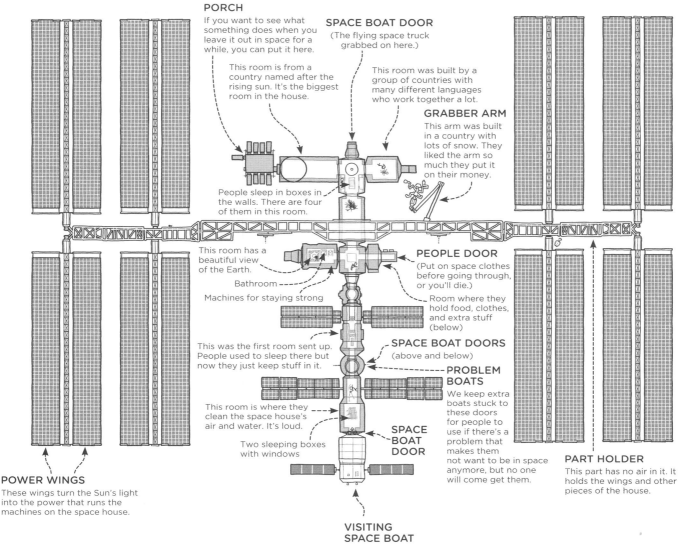

PORCH
If you want to see what something does when you leave it out in space for a while, you can put it here.

This room is from a country named after the rising sun. It's the biggest room in the house.

SPACE BOAT DOOR
(The flying space truck grabbed on here.)

This room was built by a group of countries with many different languages who work together a lot.

GRABBER ARM
This arm was built in a country with lots of snow. They liked the arm so much they put it on their money.

People sleep in boxes in the walls. There are four of them in this room.

This room has a beautiful view of the Earth.

Bathroom

Machines for staying strong

This was the first room sent up. People used to sleep there but now they just keep stuff in it.

PEOPLE DOOR
(Put on space clothes before going through, or you'll die.)

Room where they hold food, clothes, and extra stuff (below)

SPACE BOAT DOORS
(above and below)

PROBLEM BOATS
We keep extra boats stuck to these doors for people to use if there's a problem that makes them not want to be in space anymore, but no one will come get them.

This room is where they clean the space house's air and water. It's loud.

SPACE BOAT DOOR

Two sleeping boxes with windows

POWER WINGS
These wings turn the Sun's light into the power that runs the machines on the space house.

PART HOLDER
This part has no air in it. It holds the wings and other pieces of the house.

VISITING SPACE BOAT

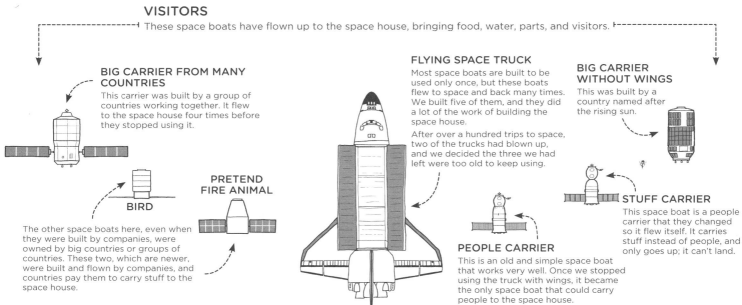

VISITORS
These space boats have flown up to the space house, bringing food, water, parts, and visitors.

BIG CARRIER FROM MANY COUNTRIES
This carrier was built by a group of countries working together. It flew to the space house four times before they stopped using it.

BIRD

PRETEND FIRE ANIMAL

The other space boats here, even when they were built by companies, were owned by big countries or groups of countries. These two, which are newer, were built and flown by companies, and countries pay them to carry stuff to the space house.

FLYING SPACE TRUCK
Most space boats are built to be used only once, but these boats flew to space and back many times. We built five of them, and they did a lot of the work of building the space house.

After over a hundred trips to space, two of the trucks had blown up, and we decided the three we had left were too old to keep using.

PEOPLE CARRIER
This is an old and simple space boat that works very well. Once we stopped using the truck with wings, it became the only space boat that could carry people to the space house.

BIG CARRIER WITHOUT WINGS
This was built by a country named after the rising sun.

STUFF CARRIER
This space boat is a people carrier that they changed so it flew itself. It carries stuff instead of people, and only goes up; it can't land.

TINY BAGS OF WATER YOU'RE MADE OF

Everything that's alive is made of tiny bags of water. Some living things are made of just one bag of water. Those things are usually too small to see. Other things are made of a group of bags stuck together. Your body is a group of lots and lots of these bags that are working together to read this page.

These bags are full of smaller bags. Life uses lots of bags. All life is made from different kinds of water, and a bag keeps the stuff inside it from touching the stuff on the outside. By using bags, living things can keep different kinds of water in one place without it all coming together.

Some of the little bags you see here were once living things on their own. Long ago, some little green bags learned to get power from the Sun. Then they got stuck inside other bags, and those became flowers and trees. The green color of leaves comes from the children of those little green bags.

LITTLE ANIMALS
These are living things (not really "animals") that got stuck in our bags of water a long time ago, like the green things in tree leaves. Now we can't live without each other. They get food and air from our bodies and turn them into power for our bags.

SIZE
These bags are almost always too small to see. In fact, they're almost as small as the waves of light we see with:

BLUE ∿∿∿∿∿∿∿
GREEN ～～～～～
RED 〜〜〜〜

BAG FILLER
This machine fills little bags with stuff and then sends them out into the water. Some stuff gets sent out of the big bag to another part of your body.

The machine also fills bags with death water, marking them very carefully before sending them out so they don't get used in the wrong place.

OUTSIDE WALL
The water bags that make up animals have soft walls. The bags in trees and flowers, which don't need to move around as much as us, have a less soft outside layer.

GETTING IN AND OUT
Some things can go through the bag's wall on their own. Other things can only go through if the bag helps them, either by letting them through an opening, or by making part of the wall into a new bag to hold them.

STOP IT!

THINGS THAT MAKE YOU SICK
These tiny things can get into your bags and take control of them. When they do that, they use the bag to build more of them.

When the kind shown here gets into you, your body gets hot, your legs hurt, and you have to lie down. Your whole body feels bad, and it makes you hate everything. You feel like you're going to die but usually don't.

We say all life is made of bags, but these things aren't. They also can't make more of themselves; they have to get a bag to make them. So we don't know if it makes sense to say they're "alive." They're more like an idea that spreads itself.

EMPTY POCKETS
This part of the bag has pockets to hold stuff that it might need later. It also makes a few things.

One of the things it makes is that stuff that helps your arms and legs get stronger. Sometimes, people who want to run or ride fast will put bottles of that stuff into their body and then lie about it.

INFORMATION
The information for how to make different body parts is stored here.

READERS →
These machines read the information about how to make parts and write it on little notes, then send them out through the holes in the wall.

MACHINE MAKER
This part makes the little machines that sit outside the control area.

CONTROL AREA
This area in the middle holds information about how to make the different parts of your body. It writes this information in notes and sends them out into the bag.

Bags make more bags by breaking in half. When this happens, the control area also breaks in half, and each half gets a full set of the bag's information.

Not all bags have these control areas. The bags in human blood don't (which means blood can't grow) but the bags in bird blood do.

This control area may have once been a living thing on its own, just like the green things in leaves.

CONTROL AREA HOLES
Notes and workers go out through these openings.

STRANGE BOXES
There are lots of these little boxes in our water bags. We don't know what they do.

BAGS OF DEATH WATER
These little bags are full of a kind of water that breaks things into tiny pieces. If something is put inside them, the water breaks it down into whatever it's made of.

If something goes wrong, these little bags tear open and all their bad water falls out. That makes the whole bag around it fall to pieces and die.

"Bags falling to pieces" sounds bad, since bags are what you're made of. But if a bag was having problems, it could hurt you. The death water helps clear it away so your body can make a new one.

BAG SHAPERS
The space between bag parts is full of lots of very thin hair-like lines. These are like bones for the bag; they help hold its shape, and do some other things.

Some of these shapers also have holes down the middle, and can carry things from one part of the bag to another.

LITTLE BUILDERS
This area is covered in little building machines that build new parts for the bag. The builders sit just outside the control area, reading the notes from inside that tell them what to build.

After the builder makes a part, the part falls away into the bag. Each part has a job to do. Maybe its job is to tell another part it's time to stop working. Maybe its job is to turn one kind of part into another. Maybe another part do something different. Or maybe it has a job, but waits until it sees *another* part before it starts working.

The strange thing is, no one tells the part where to go. It just falls out into the room with all the other parts, and hangs around until it runs into whatever part it's supposed to grab. (Or until another part grabs *it!*) This sounds strange, and it is! There are so many parts, and they're all grabbing each other and stopping each other and helping each other.

The insides of these bags are harder to understand than almost anything else in the world.

HEAVY METAL POWER BUILDING

These buildings use special kinds of hard-to-find heavy metal to make power.

Some of the metals they use can be found in the ground, but only in a few places. Other kinds can be made by people—but only with the help of a power building that's already running.

These metals make heat all the time, even when they're just sitting. They make two kinds of heat: normal heat—like heat from a fire—and a different, special kind of heat.

This special heat is like light that you can't see. (At least, you can't see it most of the time. If there's a whole lot of it, enough to kill you quickly, you can see it. It looks blue.)

Normal heat can burn you, but the special heat from these metals can burn you in a different way. If you spend too much time near this heat, your body can start growing wrong. Some of the first people who tried to learn about these metals died that way.

The special heat is made when tiny pieces of the metal break down. This lets out a lot of heat, far more than any normal fire could. But.for many kinds of metal, it happens very slowly. A piece of metal as old as the Earth might be only half broken down by now.

Within the last hundred years, we learned something very strange: When some of these metals feel special heat, they break down faster.

If you put a piece of this metal close to another piece, it will make heat, which will make the other piece break down faster and make more heat.

If you put too much of the metal together like this, it gets hotter and hotter so fast that it can all break down at once, letting out all its heat in less than a second. This is how a small machine can burn an entire city.

To make power, people try to put pieces of this metal close enough together that they make heat fast, but not so close that they go out of control and blow up. This is very hard, but there is so much heat and power stored in this metal that some people have wanted to try anyway.

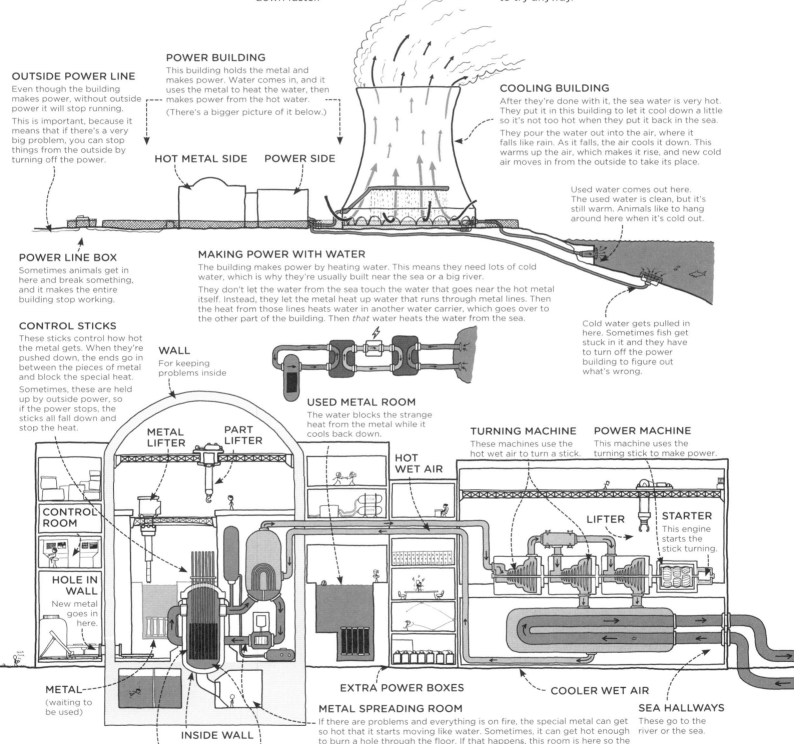

OUTSIDE POWER LINE
Even though the building makes power, without outside power it will stop running.

This is important, because it means that if there's a very big problem, you can stop things from the outside by turning off the power.

POWER BUILDING
This building holds the metal and makes power. Water comes in, and it uses the metal to heat the water, then makes power from the hot water.

(There's a bigger picture of it below.)

HOT METAL SIDE   POWER SIDE

COOLING BUILDING
After they're done with it, the sea water is very hot. They put it in this building to let it cool down a little so it's not too hot when they put it back in the sea.

They pour the water out into the air, where it falls like rain. As it falls, the air cools it down. This warms up the air, which makes it rise, and new cold air moves in from the outside to take its place.

Used water comes out here. The used water is clean, but it's still warm. Animals like to hang around here when it's cold out.

POWER LINE BOX
Sometimes animals get in here and break something, and it makes the entire building stop working.

MAKING POWER WITH WATER
The building makes power by heating water. This means they need lots of cold water, which is why they're usually built near the sea or a big river.

They don't let the water from the sea touch the water that goes near the hot metal itself. Instead, they let the metal heat up water that runs through metal lines. Then the heat from those lines heats water in another water carrier, which goes over to the other part of the building. Then *that* water heats the water from the sea.

Cold water gets pulled in here. Sometimes fish get stuck in it and they have to turn off the power building to figure out what's wrong.

CONTROL STICKS
These sticks control how hot the metal gets. When they're pushed down, the ends go in between the pieces of metal and block the special heat.

Sometimes, these are held up by outside power, so if the power stops, the sticks all fall down and stop the heat.

WALL
For keeping problems inside

USED METAL ROOM
The water blocks the strange heat from the metal while it cools back down.

HOT WET AIR

TURNING MACHINE
These machines use the hot wet air to turn a stick.

POWER MACHINE
This machine uses the turning stick to make power.

METAL LIFTER   **PART LIFTER**

CONTROL ROOM

HOLE IN WALL
New metal goes in here.

LIFTER

STARTER
This engine starts the stick turning.

METAL
(waiting to be used)

INSIDE WALL

HOT METAL   HOT WATER

EXTRA POWER BOXES

METAL SPREADING ROOM
If there are problems and everything is on fire, the special metal can get so hot that it starts moving like water. Sometimes, it can get hot enough to burn a hole through the floor. If that happens, this room is here so the watery metal can fall down and then spread out over the floor.

It's good if the metal can spread out, since when it's all close together, it keeps making itself hotter. If this room ever gets used, it means everything has gone very, very wrong.

COOLER WET AIR

SEA HALLWAYS
These go to the river or the sea.

RED WORLD SPACE CAR

This is a space car that drives around on the red world near Earth. Humans have never been to the red world, but we've sent four cars, along with lots of space ships flying around taking pictures from high above. This car is the biggest one we've sent so far—as big as a normal car on Earth.

The cars we've sent there are looking for water, because if there's water, there might be life. Right now, there's only a little water there, and it's so cold that it's all hidden away in the ground as snow. But it wasn't always that way!

By looking at the red world's rocks, our cars have learned something really cool: A long time ago, when the red world was young, it had seas.

We don't think there's life on the red world today. We haven't found any so far, and it's very cold and dry, with very thin air, so water can't last long on the ground before it turns to ice or air.

But if there used to be seas, then maybe there used to be animals, too. On Earth, when animals die, sometimes parts of their bodies turn to a kind of stone. If there were animals on the red world, maybe we can find the stones they left behind.

If we find that there was life on the red world, it will be one of the most important things we've ever learned—because if there was life on the red world, it means there's probably life in lots of places.

We now know that most of the stars in the sky have worlds around them, but we don't know if there's life on those worlds. We know there's life on our own world, but that doesn't tell us whether life is normal or not. Maybe life is a very strange thing that got started only once, and none of the other worlds have anyone on them to wonder about this question.

But if we learn that life started on the red world, too, it means that life probably gets started on new worlds all the time, and probably also got started around many of those other stars.

If our space car finds signs of life in the red world's stones, it means we are not alone.

LANDING THE SPACE CAR

Because this car was so heavy, it was hard for us to make it slow down enough to land without breaking. We could hang a big sheet behind it to slow it down, but it's so heavy—and the air there so thin—that the sheet wouldn't be able to slow it down enough.

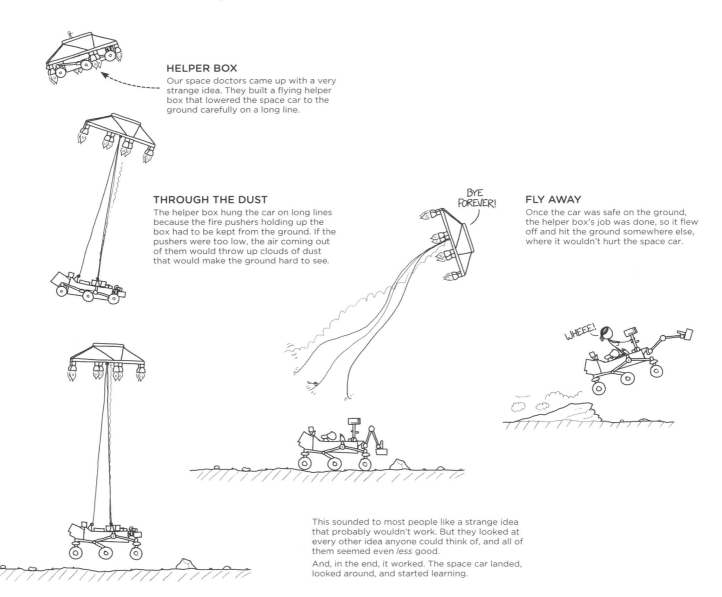

HELPER BOX

Our space doctors came up with a very strange idea. They built a flying helper box that lowered the space car to the ground carefully on a long line.

THROUGH THE DUST

The helper box hung the car on long lines because the fire pushers holding up the box had to be kept from the ground. If the pushers were too low, the air coming out of them would throw up clouds of dust that would make the ground hard to see.

FLY AWAY

Once the car was safe on the ground, the helper box's job was done, so it flew off and hit the ground somewhere else, where it wouldn't hurt the space car.

BYE FOREVER!

WHEEE!

This sounded to most people like a strange idea that probably wouldn't work. But they looked at every other idea anyone could think of, and all of them seemed even *less* good.

And, in the end, it worked. The space car landed, looked around, and started learning.

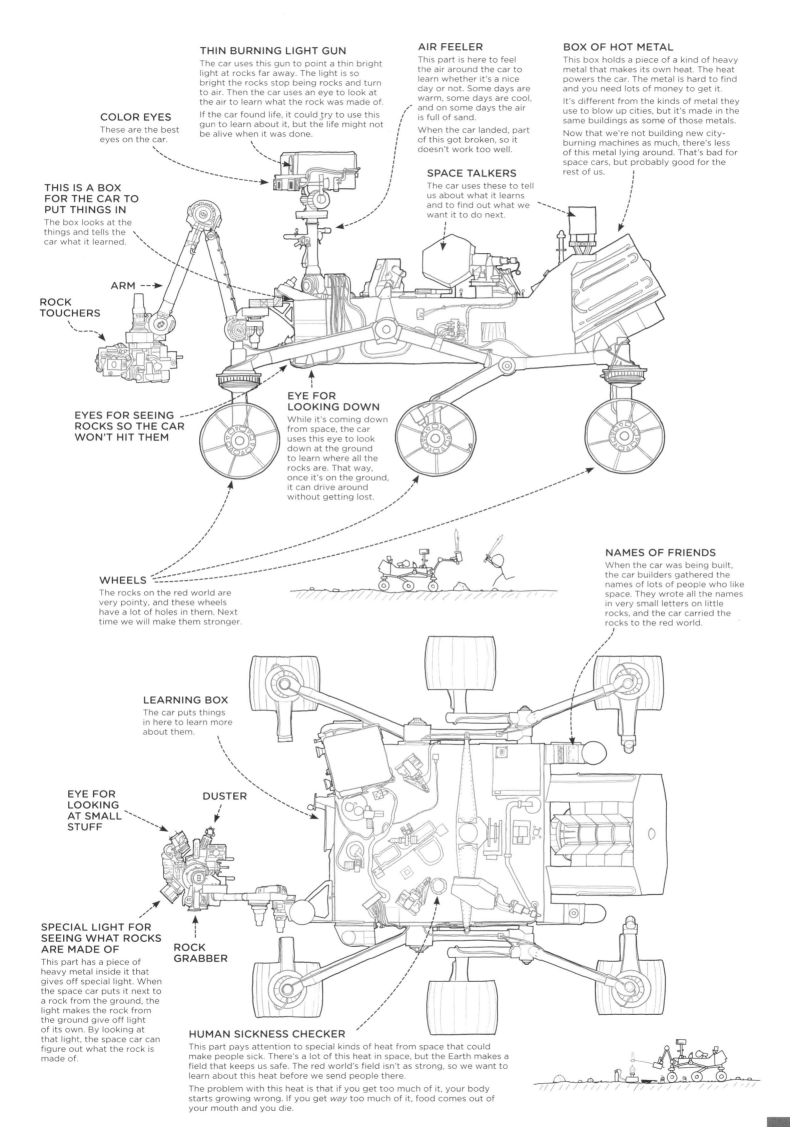

THIN BURNING LIGHT GUN

The car uses this gun to point a thin bright light at rocks far away. The light is so bright the rocks stop being rocks and turn to air. Then the car uses an eye to look at the air to learn what the rock was made of.

If the car found life, it could try to use this gun to learn about it, but the life might not be alive when it was done.

AIR FEELER

This part is here to feel the air around the car to learn whether it's a nice day or not. Some days are warm, some days are cool, and on some days the air is full of sand.

When the car landed, part of this got broken, so it doesn't work too well.

BOX OF HOT METAL

This box holds a piece of a kind of heavy metal that makes its own heat. The heat powers the car. The metal is hard to find and you need lots of money to get it.

It's different from the kinds of metal they use to blow up cities, but it's made in the same buildings as some of those metals.

Now that we're not building new city-burning machines as much, there's less of this metal lying around. That's bad for space cars, but probably good for the rest of us.

COLOR EYES

These are the best eyes on the car.

THIS IS A BOX FOR THE CAR TO PUT THINGS IN

The box looks at the things and tells the car what it learned.

SPACE TALKERS

The car uses these to tell us about what it learns and to find out what we want it to do next.

ARM

ROCK TOUCHERS

EYES FOR SEEING ROCKS SO THE CAR WON'T HIT THEM

EYE FOR LOOKING DOWN

While it's coming down from space, the car uses this eye to look down at the ground to learn where all the rocks are. That way, once it's on the ground, it can drive around without getting lost.

WHEELS

The rocks on the red world are very pointy, and these wheels have a lot of holes in them. Next time we will make them stronger.

NAMES OF FRIENDS

When the car was being built, the car builders gathered the names of lots of people who like space. They wrote all the names in very small letters on little rocks, and the car carried the rocks to the red world.

LEARNING BOX

The car puts things in here to learn more about them.

EYE FOR LOOKING AT SMALL STUFF

DUSTER

SPECIAL LIGHT FOR SEEING WHAT ROCKS ARE MADE OF

This part has a piece of heavy metal inside it that gives off special light. When the space car puts it next to a rock from the ground, the light makes the rock from the ground give off light of its own. By looking at that light, the space car can figure out what the rock is made of.

ROCK GRABBER

HUMAN SICKNESS CHECKER

This part pays attention to special kinds of heat from space that could make people sick. There's a lot of this heat in space, but the Earth makes a field that keeps us safe. The red world's field isn't as strong, so we want to learn about this heat before we send people there.

The problem with this heat is that if you get too much of it, your body starts growing wrong. If you get *way* too much of it, food comes out of your mouth and you die.

BAGS OF STUFF INSIDE YOU

This is a map of some of the different bags in your body and how they join together.

It doesn't show what they're really shaped like or how they're pushed together inside your body.

In that way, it's kind of like the colored maps in cities that tell you where trains go—it shows how the places are joined together, but not what they're shaped like or how far away from each other they are.

There are lots of important parts of your body that aren't shown on this map. But that's okay; a body has too many parts to show on *any* map.

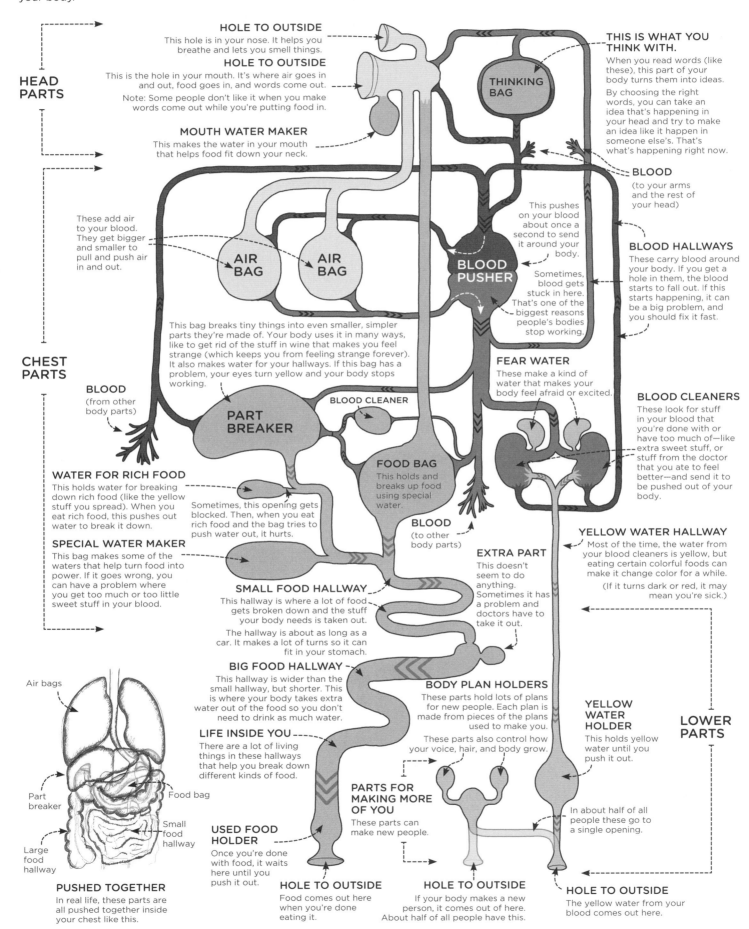

HEAD PARTS

CHEST PARTS

LOWER PARTS

HOLE TO OUTSIDE
This hole is in your nose. It helps you breathe and lets you smell things.

HOLE TO OUTSIDE
This is the hole in your mouth. It's where air goes in and out, food goes in, and words come out.
Note: Some people don't like it when you make words come out while you're putting food in.

MOUTH WATER MAKER
This makes the water in your mouth that helps food fit down your neck.

THINKING BAG

THIS IS WHAT YOU THINK WITH.
When you read words (like these), this part of your body turns them into ideas.
By choosing the right words, you can take an idea that's happening in your head and try to make an idea like it happen in someone else's. That's what's happening right now.

BLOOD
(to your arms and the rest of your head)

These add air to your blood. They get bigger and smaller to pull and push air in and out.

AIR BAG

AIR BAG

BLOOD PUSHER
This pushes on your blood about once a second to send it around your body.
Sometimes, blood gets stuck in here. That's one of the biggest reasons people's bodies stop working.

BLOOD HALLWAYS
These carry blood around your body. If you get a hole in them, the blood starts to fall out. If this starts happening, it can be a big problem, and you should fix it fast.

This bag breaks tiny things into even smaller, simpler parts they're made of. Your body uses it in many ways, like to get rid of the stuff in wine that makes you feel strange (which keeps you from feeling strange forever). It also makes water for your hallways. If this bag has a problem, your eyes turn yellow and your body stops working.

FEAR WATER
These make a kind of water that makes your body feel afraid or excited.

BLOOD
(from other body parts)

BLOOD CLEANER

PART BREAKER

BLOOD CLEANERS
These look for stuff in your blood that you're done with or have too much of—like extra sweet stuff, or stuff from the doctor that you ate to feel better—and send it to be pushed out of your body.

WATER FOR RICH FOOD
This holds water for breaking down rich food (like the yellow stuff you spread). When you eat rich food, this pushes out water to break it down.

FOOD BAG
This holds and breaks up food using special water.

Sometimes, this opening gets blocked. Then, when you eat rich food and the bag tries to push water out, it hurts.

BLOOD
(to other body parts)

EXTRA PART
This doesn't seem to do anything. Sometimes it has a problem and doctors have to take it out.

YELLOW WATER HALLWAY
Most of the time, the water from your blood cleaners is yellow, but eating certain colorful foods can make it change color for a while.
(If it turns dark or red, it may mean you're sick.)

SPECIAL WATER MAKER
This bag makes some of the waters that help turn food into power. If it goes wrong, you can have a problem where you get too much or too little sweet stuff in your blood.

SMALL FOOD HALLWAY
This hallway is where a lot of food gets broken down and the stuff your body needs is taken out.
The hallway is about as long as a car. It makes a lot of turns so it can fit in your stomach.

BIG FOOD HALLWAY
This hallway is wider than the small hallway, but shorter. This is where your body takes extra water out of the food so you don't need to drink as much water.

BODY PLAN HOLDERS
These parts hold lots of plans for new people. Each plan is made from pieces of the plans used to make you.
These parts also control how your voice, hair, and body grow.

YELLOW WATER HOLDER
This holds yellow water until you push it out.

Air bags

Part breaker

Food bag

Large food hallway

Small food hallway

LIFE INSIDE YOU
There are a lot of living things in these hallways that help you break down different kinds of food.

PARTS FOR MAKING MORE OF YOU
These parts can make new people.

In about half of all people these go to a single opening.

PUSHED TOGETHER
In real life, these parts are all pushed together inside your chest like this.

USED FOOD HOLDER
Once you're done with food, it waits here until you push it out.

HOLE TO OUTSIDE
Food comes out here when you're done eating it.

HOLE TO OUTSIDE
If your body makes a new person, it comes out of here. About half of all people have this.

HOLE TO OUTSIDE
The yellow water from your blood comes out here.

BOXES THAT MAKE CLOTHES SMELL BETTER

Clothes don't stay clean for long. Pieces of dust and dirt stick to them, and they get coated in that clear stuff that comes off your skin. If clothes get wet for too long, things can grow on them, which makes them smell bad.

This box holds two machines that clean clothes. The bottom one washes them with water, and the top one dries them.

DRYER

DUST CATCHER
When the air blows through the clothes, it carries away bits of dust and tiny pieces of the clothes. This thing catches the dust so it doesn't gather somewhere else in the house.

When the catcher fills up, you have to clean it out, because if it's full of dust, it blocks the air. This keeps the machine from drying—and the dust is easy to burn, so it can help a fire start.

MMMM...

For some reason, some people really like pulling the sheets of dust out of the catcher.

HEATER
This works the same way as a hair dryer. Power runs through metal lines. This makes the metal hot for the same reason lights get hot. Then the air blows over the metal.

OUTSIDE HOLE
This carries the hot air out of the house.

On a cold day, sometimes you walk past these holes when the machine is running, and the warm air feels nice on your face and smells like clean clothes.

Hot air going out

Line that turns the clothes box

DOOR

Hot air coming in

BLOWER

Wheels the clothes box sits on so it can turn

WHY CLEANING IS HARD
You can wash away some kinds of dirt using water, because dirt sticks to water and gets carried away by it. But other things that make clothes dirty, like some stuff your skin makes, don't stick to water.

To get rid of stuff that doesn't stick to water, we use special cleaning stuff. When you put this stuff in with your clothes, it sticks to the stuff that makes clothes dirty, but it *also* sticks to the water around it. Then, when you shake everything, the water pulls the dirt free.

Water

Clothes

Dirt

Cleaning stuff

CONTROLS
You use these to decide how clean you want your clothes to be and how careful you want the machine to be with them.

WASHING		DRYING	
Water heat	Shaking hardness	Air heat	Running time

Hot water cleans better but can also wash out colors.

Hard shaking cleans better but can tear clothes.

Hot air dries better but hurts clothes more.

Longer time dries better but hurts clothes more.

POWER SPINNER
This spins the clothes holder so the clothes turn over. If it didn't, only the top surface of the clothes would get dry. It also turns the air blower, which pushes air into the hot air box.

SPINNING REALLY FAST
It's hard to get water out of clothes. To do it, the cup spins really fast. The edge of the cup goes about as fast as the fastest horse.

This pushes the clothes against the side of the cup, and it makes the water fall out of the clothes and into holes in the cup wall. Then the water falls to the bottom, and the water mover pulls it out to get rid of it.

SOFT HOLDERS
Because the clothes cup spins so fast, it's hard to keep it from shaking, making loud noises, or breaking.

To make the cup quieter and keep it from breaking, they hang it on holders that can stretch. This lets it move around a little, which makes it quieter. (It's just like when someone calls you and your phone shakes; it's much louder if it's sitting on a hard table than if it's on a soft bed.)

Letting the cup move around makes it quieter, but if all the clothes end up on one side of the cup, it can move around *too* much. Then the machine starts making a loud noise as it shakes. Most machines can tell when this is happening and turn themselves off; if they didn't, they could shake themselves to pieces.

DOOR
You usually pour the cleaning stuff in here with the clothes, though some machines have a second, smaller door for this.

CLOTHES CUP
This fills with water to clean the clothes.

WASHER

CLOTHES PUSHER
This thing turns one way and then the other to move the clothes up and down, to make sure they all get covered in water and cleaning stuff.

WATER COMING IN

There are two layers to the cup. The inside layer can spin, and has holes to let water through to the outside layer so the water mover can pull it out.

WATER GOING OUT

HOUSE WATER
These two lines bring hot and cold water from the wall of your house.

POWER LINE
The washer doesn't take too much power to run, but the dryer takes a lot.

WATER MOVER
This machine pulls water from the bottom of the big cup and sends it away into your house's system for getting rid of used water.

SPIN CHANGER
This machine lets the power spinner turn the clothes cup fast—to pull water out—or move the clothes pusher slowly, to shake the clothes around in the water.

POWER SPINNER
This turns the clothes cup and the clothes pusher in the middle. It also runs the water mover.

WAIT
Why is this in your house?

EARTH'S SURFACE

These maps show the Earth's surface. The Earth's surface is special, as far as we know. It's the only place where we've found seas of water, and the only place where the land is made of sheets of rock that move around. There are a lot of interesting things here. These maps show where some of them are.

Earth is a round ball, so to fit its surface on a page, it has to be stretched out. This changes the shapes and sizes of some areas. On this map, it makes the land at the top and the bottom look much bigger than it really is, and some of the places near the sides look stretched out.

There's no way around this problem. Every paper map of a round world is wrong about size, shape, or the direction from one place to another. The shape chosen for this map tries to keep all these things in mind, not stretching any one part too much or making any area look too wrong.

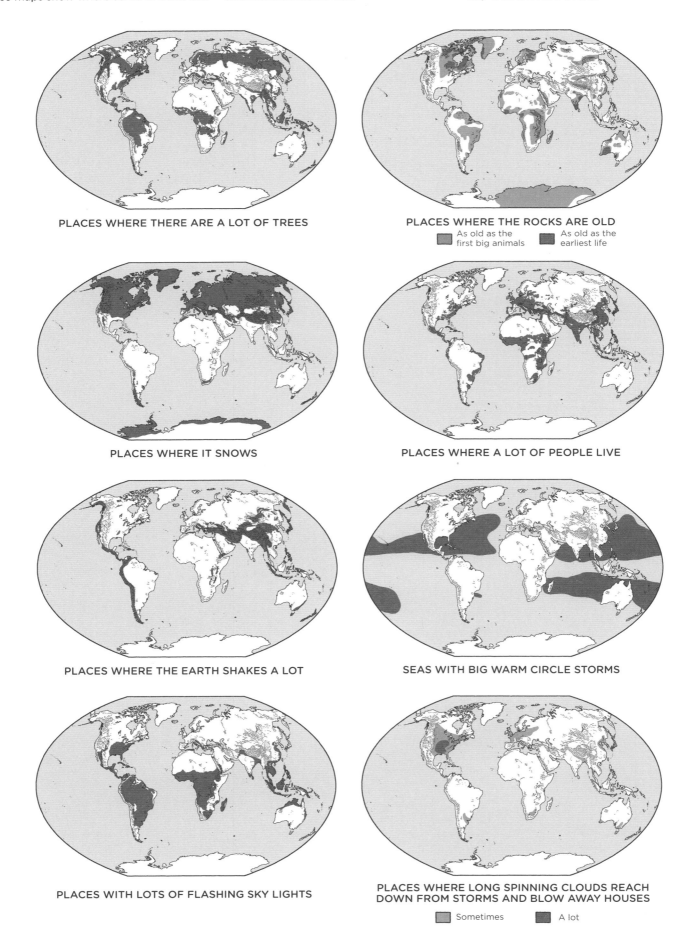

PLACES WHERE THERE ARE A LOT OF TREES

PLACES WHERE THE ROCKS ARE OLD

As old as the first big animals   As old as the earliest life

PLACES WHERE IT SNOWS

PLACES WHERE A LOT OF PEOPLE LIVE

PLACES WHERE THE EARTH SHAKES A LOT

SEAS WITH BIG WARM CIRCLE STORMS

PLACES WITH LOTS OF FLASHING SKY LIGHTS

PLACES WHERE LONG SPINNING CLOUDS REACH DOWN FROM STORMS AND BLOW AWAY HOUSES

Sometimes   A lot

SKY BOAT WITH TURNING WINGS

Normal sky boats have to go fast so the air hits their wings hard enough to hold them up. If they fly too slow, they fall. (Sometimes falling makes them go fast enough to fix it!)

This boat works just like those sky boats, but it uses a cool idea: Instead of the whole boat going fast, only its *wings* go fast. The rest of the boat can go as slowly as it wants—or even stop and sit in one place in the sky.

If a normal sky boat had wings that went faster than its body, the wings would fly away, but this boat's wings go in a circle. That keeps them near enough to hold on to while still going around fast enough to fly.

TURNING WINGS
These are a lot like the wings on normal sky boats, but they go around instead of forward.

POINTING WINGS
These wings keep the boat pointed straight. (The pusher next to it helps with that, too.)

POWER SPINNER
This is a fire water machine that works like a sky boat pusher, but this kind of boat uses all the spinner's power for turning the stick holding the turning wings. Hot air comes out, like in a sky boat pusher, but it doesn't push on anything.

SPIN CHANGER
To work best, the power spinner needs to turn very fast, but the wings can't go around that fast. The spin changer uses wheels with teeth to make the wings turn more slowly than the power spinner.

If the boat didn't have this box, the wings would turn the same number of times each second as the power spinner, and the ends of the wings would go a lot faster than sound. That would make them stop working, and probably break off.

BOAT HANGER
The boat hangs under the wings, and this piece of metal holds them together.

WING CONTROLLER
This is a machine for making small changes to how the wings are turned as they swing around, which changes how they push on the air. (It's a little confusing—you can read more about how it works below.)

COOL AIR COMING IN

HOT AIR GOING OUT
(but not helping to push)

CONTROL LINES
These use water to push the wing controls around (see below).

WINDOW

TURNING STICK
This stick turns the end pusher.

RADIO STICK
This line feels radio waves.

END BLOWER
When the boat's wings turn, it pushes the boat in the other direction. This end blower pushes back, keeping the boat from spinning.

AH, YES, THEY GO THE OTHER WAY IN THE SOUTHERN HALF OF THE WORLD BECAUSE OF THE EARTH'S SPIN.

DID YOUR MIND FALL OUT OF YOUR HEAD?

In a few countries, sky boat wings turn right. Most turn left.

BOTTOM WINDOW
When your boat can land straight down, it helps to be able to *look* down.

LAND FEET
Some sky boats use these instead of wheels, since they often land in places like grass or dirt where wheels get stuck.

HOW THESE SKY BOATS GO

A sky boat's wings can turn so that they're going straight into the wind, or so that the wind is pushing them up. If they turn straight into the wind, they don't lift the sky boat.

To go forward, a sky boat controls its wings so the one in the front goes straight into the wind, and the one in the back turns so it gets pushed up.

This makes the the sky boat lean forward. Before, the wings were just pulling it up, but after it leans forward a little, they pull it a little forward, too. The more it leans, the faster it goes forward. If it leans too far, the wings will *only* pull it forward—not up. This causes bad problems.

WAIT, HOW DOES THAT WORK?

You might wonder how a sky boat can turn its front wing and not its back wing, since the wings keep changing places.

The answer is that some people who were very good at machines figured out how to make a machine that turns the wings up and down as they go around.

Lifting a lot   Not lifting

Rings
Sticks
Turning
Not turning

The edge of each wing has a stick on it, and these sticks run to a ring. This ring spins with the wings and turning stick, and sits on top of another ring that *doesn't* spin.

To go forward, the driver uses a control to turn the bottom ring, which turns the spinning top ring, too, so it's higher on one side. When a wing is on that side, its stick pushes its back edge up, so it goes straight into the wind and doesn't lift at all. When it goes around to the other side, its back edge gets pulled down, making it lift a lot.

HOW HIGH CAN A SKY BOAT GO?

A sky boat with turning wings needs more air to push against than a normal sky boat. Up high, where normal sky boats fly, the air is thin because it's closer to space. Very few turning-wing sky boats can fly to the tops of the tallest mountains, but normal sky boats can fly over them with no problems.

Normal sky boat

Turning-wing sky boat

But a turning-wing sky boat can still go farther above the sea than most under-water boats can go below it.

Boat that goes under the sea

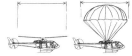

BENDING WINGS
Sometimes, sky boat wings bend down while they're sitting on the ground. This might look like it would cause some problems, but it's normal! Wings that bend a little can be easier to control, and when they're swinging around, the force of the swing stretches them out straight.

WHAT IF EVERYTHING BREAKS?

If a normal sky boat stops working, it can keep going, slowly moving down from slowing down too much. It turns out that even when it's not flying forward, a turning-wing sky boat can do the same thing!

Even though they're thin, turning wings can slow a falling boat almost as much as a big sheet.

When the engine stops, it lets go of the turning stick so the wings can spin. If the wings are turned in the right direction, the air going through them will make them spin faster, and the air rushing past pushes up on them, slowing the fall.

It may seem strange that turning the wings could help push up on the boat when there's no *power* turning them. But you may have seen this kind of free-turning lift happen without knowing it, because trees use it.

Trees make babies by dropping tiny wooden tree eggs on the ground. To help the trees spread farther, some trees put small leaf wings on their eggs to slow their fall so the wind can blow them. The wings aren't very big, so they aren't able to slow the eggs much—but they *turn*. That lets them fall very slowly and blow very far.

So don't worry if your sky boat turns off. It can still fly, just like a tiny spinning leaf, carrying you to the ground alive and safe.

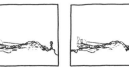

THE US'S LAWS OF THE LAND

This country was started when a group of people who were part of another country decided to break away and start their own country. They wrote down a small set of laws to be the ground that the new country—and its many future laws—would be built on.

More than two hundred years later, those laws, with some changes, are still being followed, learned about, and understood in new ways.

BEFORE WE START

Hi; we're the people in these little countries called "states," and we want to get together into a country. We want to make everything nice and quiet, keep anyone from hurting us, and make sure our kids will be free. That's why we're making a country. Here are its laws:

BOOK ONE: The Law Makers

Part One: Laws are made by a group called Law Makers. There are two rooms of Law Makers: the House and the Serious Room.

Part Two: The people pick Law Makers to send to the House for two years at a time. Bigger states get to have more people in the House. Oh, and the country needs to count its people sometimes so it can figure out how many chairs the room needs.

People used to take other people from far-away countries, bring them across the sea, and force them to work for their whole lives without getting paid. There was a part of the law here that said that when you're counting people, you should only count a forced worker as part of a person. A while after this was written, we fought a war over whether people should be allowed to own other people like that. The side that said "yes" lost, and we crossed this part out.

Part Three: Every state sends two Law Makers to the Serious Room for six years at a time. They can't be too young.

Part Four: States make the laws about where and how people get together to pick leaders and decide what the country should do.

Part Five: When the Law Makers get together, they should write down what they talk about.

Part Six: Law Makers get paid. They can't get in trouble for what they say at work, but they also can't do any other job for the country while they're Law Makers.

Part Seven: If the Law Makers have an idea for a new law, and more than half the people in both rooms say they like it, they send the idea to the country's leader to be made into a law. If the leader doesn't like the idea, the Law Makers can still make it a law, but it takes more of them.

Part Eight: Law Makers are allowed to take money from people, but only sometimes, and they can't just take it all from one person or anything like that. They're allowed to use the money to build certain kinds of things, like letter boxes and boats with guns on them. They can get people in trouble for a few things, like stealing boats (even if they do it far away) or making pretend money and telling people it's real.

Part After Eight: There are lots of things the Law Makers aren't allowed to do. They can't make up laws to lock someone up for something they already did, or give some people special names that mean they're more important to the country than others.

This part also had some laws about buying and selling people that we changed after the war.

Part Ten: There are some things the country can do that the states can't, like creating money or starting wars. The states also can't take money from other states or put guns on boats.

BOOK TWO: The leaders

Part One: Every four years, the people in the country pick who should lead them. They pick a First Leader, who is the head of the country, and a Second Leader, who isn't. If the First Leader leaves or gets fired, the Second Leader takes over the work the First Leader was doing. The states get to choose the leaders by a point system where each state gets one point for each Law Maker it has.

You may notice that this doesn't actually say whether the Second Leader becomes the First Leader or not. This made things confusing later on.

No matter how many changes people make to this system, it will probably never quite work right.

Part Two: The leader controls the people who fight for the country. The leader can also can talk to the leaders of other countries and get anyone out of trouble.

Part Three: Now and then, the First Leader should let the Law Makers know how everything is going, and suggest some ideas.

Part Four: The Law Makers can fire the First Leader, but only for doing something really bad, like becoming leader of another country at the same time and having them attack us, or stealing the country's money and going to live on a boat.

BOOK THREE: The law deciders

Part One: There's a group of people called the Top Law Deciders. They help decide whether laws have been broken. The country can set up other groups of law deciders, too, but they're not as important as the Top Law Deciders.

Part Two: The Top Law Deciders only decide certain kinds of law fights, like if the leaders of another country send someone here and they get in a fight, or when someone has a law fight with a state. The rest of the time, they can only step into certain law fights, and only when another decider has decided something and the people in the law fight don't agree.

Part Three: "Turning against the country" can only mean a few very clear things: fighting us, joining a group that's fighting us, or helping a group that's fighting us. To prove someone has turned against the country, two people need to say they saw it, or the person has to have admitted it in a deciding room. Law Makers can make turning against the country against the law, but they can't use it as a reason to do whatever they want to someone. (This had been a problem in some other countries.)

BOOK FOUR: The states

Part One: There are states, and they have to get along. When the law deciders in one state decide something, the law deciders in other states don't have to make that choice the same way, but they can't make it so the other choice doesn't count. That means that if someone gets in trouble in a state, they can't go to another state and get a law decider to tell them they're not really in trouble after all.

Part Two: You have the same rights no matter what state you're from. Also, if you get in trouble in one state and run away to another, the other state has to send you back to the first one.

Part Three: The country can add new states. The country can also own areas of land inside states (to be used for things the country needs) just like people do.

Part Four: The country promises that every state will be run by its people, and that if someone attacks—or if they have a problem and ask for help—the whole country will come and fight for them.

BOOK FIVE: Making changes

People can change these laws, but most of the Law Makers and most of the states have to agree on the change. It can't just be a little more than half—it has to be *most* of them. If the states want to make a change without the Law Makers, the states can also hold a big law party where each state comes and shares their ideas for changes, and then they decide which ones they like.

The states have never tried making a change using the "law party" idea, and no one is really sure how it would work if they did.

BOOK SIX: Everyone, listen up

These laws are important and everyone has to follow them. Also, if the country agrees on something with another country, that's important, too. Other laws are important, but less so. Anyone working for the country has to promise that they're on our side (but they never have to say anything about God).

BOOK SEVEN: Does this all count yet?

This country only becomes real if more than eight states join.

TEN CHANGES:

We made these changes right at the start, because some people said they wouldn't agree to join unless we added this stuff.

Change One: The country can't make laws about God. It also can't make laws about what people talk about, who they hang out with, or what they write about, and can't stop them from telling the leaders if they're angry about something, as long as they're not starting fights.

Change Two: Since having well-trained normal people with guns is important for keeping the country safe, no stopping people from having guns.

The way this is worded has been confusing to people over the years. Making things even worse, when it was written down for the different states and Law Makers to agree to, not all of them saw it with the same marks between the words.

Change Three: Just because someone's fighting for the country doesn't mean you have to let them stay in your house.

Change Four: The police can't go through your stuff without a good reason and a special pass from a law decider.

Change Five: The police can't do stuff to you just because they want to; they need to make it clear what you did wrong. They can never make you admit you broke a law.

Change Six: If you get in trouble, you can have a chance to fight about it in front of a group of normal people in a deciding room, and you can always have someone who knows about laws to help you if you want. If someone says you did something bad, you get to talk to them face-to-face.

Change Seven: You can have your law fight in front of a group of normal people even if you're not in trouble.

Change Eight: Police can't be mean for fun, even to bad people.

Change After Eight: People can do stuff not talked about here.

Change Ten: The country can only do the things these laws let it do. The states can do whatever.

Later, we got a little clearer about what states can and can't do.

MORE CHANGES:

These are changes we made over the next two hundred years.

Change: People can't have law fights with other states—only their own.

Change: We changed the laws for how we pick leaders.

We made this change because it started feeling strange that the people picked new leaders, yet the old ones stayed in their job for many months.

Change: We just had a big war with some states over whether it's okay to buy humans and force them to work. The side that said "no" won. No more buying humans or forcing them to work.

Change: Also, now that the war is over, we're adding a number of laws about what states can and can't do to people.

Change: Oh, and people of any skin color can help pick leaders and decide what the country will do.

. . . although it's somehow still not completely clear, even though we've tried to get it right like three or four times.

Change: The country can take some of your pay to get money for things we need.

Change: People, not a state's leaders, pick the Law Makers who will sit in the Serious Room.

Change: Let's get rid of beer and wine.

Change: People of any sex can help pick leaders and decide what the country will do.

Change: We moved up some of the days when new leaders take over for the old ones, because we have cars now and don't need to allow a few months for people to travel.

Change: Never mind about getting rid of beer and wine.

Change: You can't be the First Leader forever.

Change: People in the special town where the leaders and Law Makers live can help pick leaders and decide what the country will do, just like if they lived in a normal state.

Change: No making people pay to help decide things.

Change: We made it clearer what happens when a leader dies or leaves.

Change: Younger people can help pick leaders now.

Change: If Law Makers decide to change how much they're paid, they don't get the new pay until after the people in their state have had a chance to decide whether to fire them and pick someone else.

Some of the states agreed to this change, but then it got forgotten. Later, we found it and other states decided to pass it. The pay thing it fixes hasn't really been a problem, but it seems like an okay enough idea, so why not!

THE US'S *LAWS OF THE LAND*
(A boat)

This boat is sometimes also called "Old Metal Sides" because someone once tried to make a hole in the side but couldn't.

This boat was built to fight in wars more than two hundred years before this book was written. Even though it's old, it's still part of the country's fighting forces. That means that if someone were attacking the country with boats, and the country's leader said, "Send all our boats to that boat fight," this one would have to go, too.

Of course, that's not really going to happen, since this boat is over two hundred years old and would not help very much in a fight. Instead, the country keeps it around so they can let people visit it, help people think about the past, and teach everyone how old boats worked.

Note: There are lots of special words for things on boats. If you call this thing a "boat," people who know a lot about boats might get mad at you.

When the boat was first built, they put up a message like this around the city:

Does anyone want to help their country? Our leader told us to take this boat, which has lots of guns, and get it ready to drive around on the sea as soon as we can.

We set up a place near the bird sign on Front Street and need almost two hundred people to come help their country for a year. We'll pay ten (or more, if you're good) every month, with two months of pay ahead of time if you want. No sick people.

This is a great chance for people around here to fight for our country and get even with anyone who hurts us. Come to the place we talked about. We'll be nice to you!

Signed, the leader of the boat.

Oh, also: Someone from the fighting forces will be there looking for fighters and music players. Tall people only.

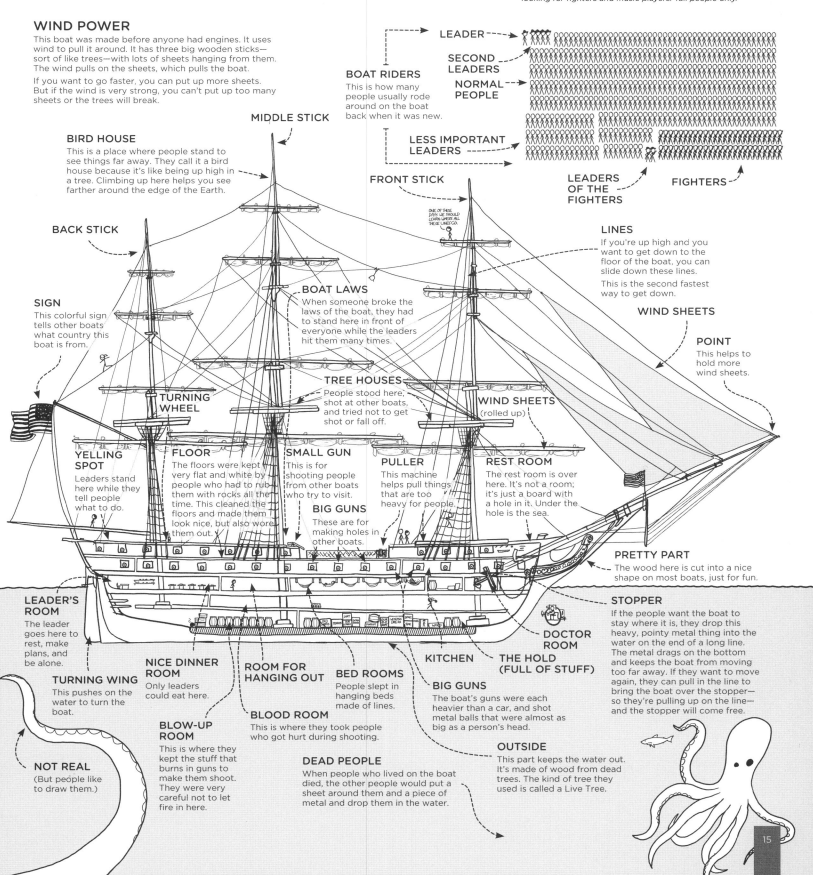

WIND POWER
This boat was made before anyone had engines. It uses wind to pull it around. It has three big wooden sticks—sort of like trees—with lots of sheets hanging from them. The wind pulls on the sheets, which pulls the boat.

If you want to go faster, you can put up more sheets. But if the wind is very strong, you can't put up too many sheets or the trees will break.

BOAT RIDERS
This is how many people usually rode around on the boat back when it was new.

LEADER
SECOND LEADERS
NORMAL PEOPLE
LESS IMPORTANT LEADERS
LEADERS OF THE FIGHTERS
FIGHTERS

MIDDLE STICK

BIRD HOUSE
This is a place where people stand to see things far away. They call it a bird house because it's like being up high in a tree. Climbing up here helps you see farther around the edge of the Earth.

BACK STICK

FRONT STICK

ONE OF THESE DAYS WE SHOULD LEARN WHERE ALL THESE LINES GO.

LINES
If you're up high and you want to get down to the floor of the boat, you can slide down these lines.

This is the second fastest way to get down.

WIND SHEETS

SIGN
This colorful sign tells other boats what country this boat is from.

BOAT LAWS
When someone broke the laws of the boat, they had to stand here in front of everyone while the leaders hit them many times.

POINT
This helps to hold more wind sheets.

TREE HOUSES
People stood here, shot at other boats, and tried not to get shot or fall off.

WIND SHEETS
(rolled up)

TURNING WHEEL

YELLING SPOT
Leaders stand here while they tell people what to do.

FLOOR
The floors were kept very flat and white by people who had to rub them with rocks all the time. This cleaned the floors and made them look nice, but also wore them out.

SMALL GUN
This is for shooting people from other boats who try to visit.

BIG GUNS
These are for making holes in other boats.

PULLER
This machine helps pull things that are too heavy for people.

REST ROOM
The rest room is over here. It's not a room; it's just a board with a hole in it. Under the hole is the sea.

PRETTY PART
The wood here is cut into a nice shape on most boats, just for fun.

LEADER'S ROOM
The leader goes here to rest, make plans, and be alone.

TURNING WING
This pushes on the water to turn the boat.

NICE DINNER ROOM
Only leaders could eat here.

ROOM FOR HANGING OUT

BED ROOMS
People slept in hanging beds made of lines.

KITCHEN

DOCTOR ROOM

THE HOLD (FULL OF STUFF)

STOPPER
If the people want the boat to stay where it is, they drop this heavy, pointy metal thing into the water on the end of a long line. The metal drags on the bottom and keeps the boat from moving too far away. If they want to move again, they can pull in the line to bring the boat over the stopper—so they're pulling up on the line—and the stopper will come free.

NOT REAL
(But people like to draw them.)

BLOW-UP ROOM
This is where they kept the stuff that burns in guns to make them shoot. They were very careful not to let fire in here.

BLOOD ROOM
This is where they took people who got hurt during shooting.

DEAD PEOPLE
When people who lived on the boat died, the other people would put a sheet around them and a piece of metal and drop them in the water.

BIG GUNS
The boat's guns were each heavier than a car, and shot metal balls that were almost as big as a person's head.

OUTSIDE
This part keeps the water out. It's made of wood from dead trees. The kind of tree they used is called a Live Tree.

FOOD-HEATING RADIO BOX

These boxes use radio waves to heat food. Radio waves push on the tiny pieces water is made of and make them go faster. When tiny pieces in something move faster, that thing gets hotter. If you send enough radio waves through water, the water heats up.

Food-heating radio boxes can heat up cold food you saved, and let you buy food that's full of ice, keep it for a long time, and then heat it and get rid of the ice. These boxes made it much easier for people to eat without spending a long time making their food.

You can also use a radio box to take fresh food (like fish) and heat it up and turn it into different kinds of food, just like you do with the other heating boxes in your kitchen. But it can be hard to use for that, so be careful, especially with food made from animals.

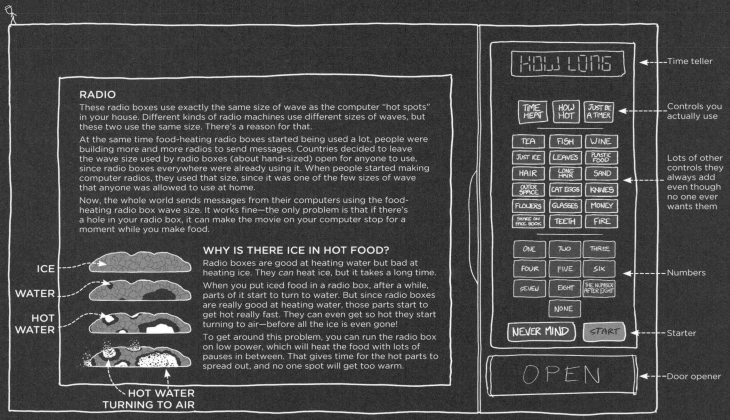

RADIO

These radio boxes use exactly the same size of wave as the computer "hot spots" in your house. Different kinds of radio machines use different sizes of waves, but these two use the same size. There's a reason for that.

At the same time food-heating radio boxes started being used a lot, people were building more and more radios to send messages. Countries decided to leave the wave size used by radio boxes (about hand-sized) open for anyone to use, since radio boxes everywhere were already using it. When people started making computer radios, they used that size, since it was one of the few sizes of wave that anyone was allowed to use at home.

Now, the whole world sends messages from their computers using the food-heating radio box wave size. It works fine—the only problem is that if there's a hole in your radio box, it can make the movie on your computer stop for a moment while you make food.

ICE
WATER
HOT WATER
HOT WATER TURNING TO AIR

WHY IS THERE ICE IN HOT FOOD?

Radio boxes are good at heating water but bad at heating ice. They *can* heat ice, but it takes a long time.

When you put iced food in a radio box, after a while, parts of it start to turn to water. But since radio boxes are really good at heating water, those parts start to get hot really fast. They can even get so hot they start turning to air—before all the ice is even gone!

To get around this problem, you can run the radio box on low power, which will heat the food with lots of pauses in between. That gives time for the hot parts to spread out, and no one spot will get too warm.

HOW LONG — Time teller

TIME HEAT | HOW HOT | JUST BE A TIMER — Controls you actually use

TEA | FISH | WINE
JUST ICE | LEAVES | PLASTIC FOOD
HAIR | LONG HAIR | SAND
OUTER SPACE | CAT EGGS | KNIVES
FLOWERS | GLASSES | MONEY
SHARE ON FACE BOOK | TEETH | FIRE
— Lots of other controls they always add even though no one ever wants them

ONE | TWO | THREE
FOUR | FIVE | SIX
SEVEN | EIGHT | THE NUMBER AFTER EIGHT
NONE
— Numbers

NEVER MIND | START — Starter

OPEN — Door opener

RADIO WAVE STOPPER
This stuff, which you see if you look inside the door, stops radio waves from getting out. They can't really hurt you—other than by slowly warming you up—but they could hurt other radios or make little flashes of light.

SPINNER
This spinner waves a metal stick to change the shape of the radio waves so the warm spots, which are places where the waves are strong, move around a little.

RADIO HALLWAY
This hallway carries the radio waves into the food box.

DOOR WATCHER
This turns off the power to the radio wave maker if the door opens so the box doesn't start to warm *you* if you open it early.

RADIO WAVE MAKER
This makes radio waves by letting power fly around through the spaces inside it. This builds up a radio wave of a certain size, like how an empty bottle plays a certain note if you blow over the hole at the top.

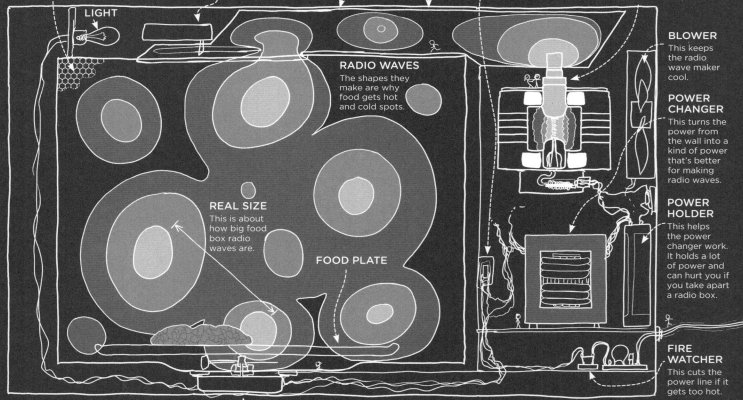

LIGHT

RADIO WAVES
The shapes they make are why food gets hot and cold spots.

REAL SIZE
This is about how big food box radio waves are.

FOOD PLATE

BLOWER
This keeps the radio wave maker cool.

POWER CHANGER
This turns the power from the wall into a kind of power that's better for making radio waves.

POWER HOLDER
This helps the power changer work. It holds a lot of power and can hurt you if you take apart a radio box.

FIRE WATCHER
This cuts the power line if it gets too hot.

This spinner turns the plate to try to give each piece of food some time in the hot areas.

SHAPE CHECKER

This machine checks whether you have a piece of metal with a certain shape. If you do, it lets go of whatever it's holding on to. People put these machines on boxes, doors, and cars to try to control who can open or use them.

What's interesting about these machines isn't really the machine itself. There are lots of different kinds that work in different ways, but they're all the same in one way: They try to put people into groups.

By checking whether someone has a piece of metal that's the right shape, this machine is really a way to try to tell whether people are who they say they are. It's an idea—about which people should be allowed to do something—brought to life in metal.

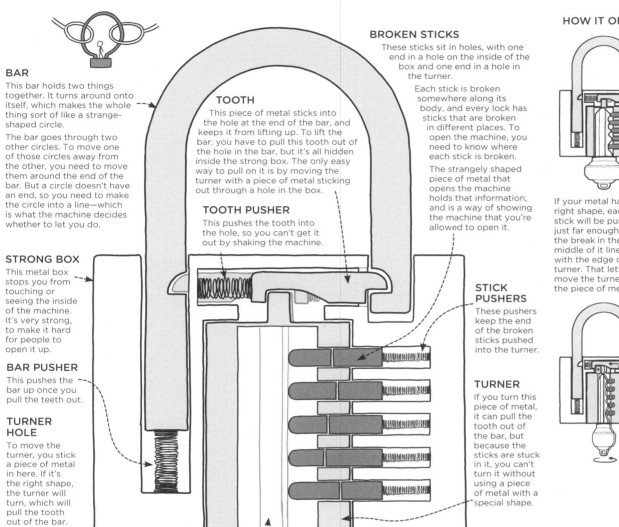

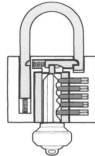

BAR
This bar holds two things together. It turns around onto itself, which makes the whole thing sort of like a strange-shaped circle.

The bar goes through two other circles. To move one of those circles away from the other, you need to move them around the end of the bar. But a circle doesn't have an end, so you need to make the circle into a line—which is what the machine decides whether to let you do.

STRONG BOX
This metal box stops you from touching or seeing the inside of the machine. It's very strong, to make it hard for people to open it up.

BAR PUSHER
This pushes the bar up once you pull the teeth out.

TURNER HOLE
To move the turner, you stick a piece of metal in here. If it's the right shape, the turner will turn, which will pull the tooth out of the bar.

TOOTH
This piece of metal sticks into the hole at the end of the bar, and keeps it from lifting up. To lift the bar, you have to pull this tooth out of the hole in the bar, but it's all hidden inside the strong box. The only easy way to pull on it is by moving the turner with a piece of metal sticking out through a hole in the box.

TOOTH PUSHER
This pushes the tooth into the hole, so you can't get it out by shaking the machine.

BROKEN STICKS
These sticks sit in holes, with one end in a hole on the inside of the box and one end in a hole in the turner.

Each stick is broken somewhere along its body, and every lock has sticks that are broken in different places. To open the machine, you need to know where each stick is broken.

The strangely shaped piece of metal that opens the machine holds that information, and is a way of showing the machine that you're allowed to open it.

STICK PUSHERS
These pushers keep the end of the broken sticks pushed into the turner.

TURNER
If you turn this piece of metal, it can pull the tooth out of the bar, but because the sticks are stuck in it, you can't turn it without using a piece of metal with a special shape.

HOW IT OPENS

To open the machine, you push a piece of metal into the hole. As it goes in, the sides of that metal push the sticks out of the way. Because of the shape of the metal, some sticks get pushed farther than others.

If your metal has the right shape, each stick will be pushed just far enough that the break in the middle of it lines up with the edge of the turner. That lets you move the turner using the piece of metal.

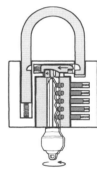

By moving the turner, you pull the tooth out of the hole in the bar. Then you can lift the bar and move the two circles away from each other.

OTHER CHECKING MACHINES
There are many other kinds of machines for checking whether someone has something (like a piece of metal or special information) and only opening if they do.

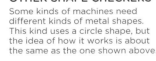

OTHER SHAPE CHECKERS
Some kinds of machines need different kinds of metal shapes. This kind uses a circle shape, but the idea of how it works is about the same as the one shown above.

NUMBER CHECKERS
Other machines check numbers instead of shapes. If you know the right numbers, you can make the machine open.

These usually work using metal wheels that turn. When the wheels are all lined up in the right way, the bar opens, but you have to know which way to line the wheels up.

One problem a lot of these machines have is that by turning them and listening and feeling very carefully, you can sometimes figure out how the wheels fit together.

And even if you can't, you can always just try all the numbers. If you're willing to wait, you can open most simple number checkers that way in a few hours.

LYING TO THE CHECKER
You can make a machine like this open even if you don't have the right shape. Here's one way to do that:

You start by pushing a thin piece of metal into the hole and gently turning.

While turning it, you reach in with a second piece of metal, and use the end to push on the broken sticks one at a time. If you lift a stick while using your other hand to turn the turner, the broken spot can get caught on the turner's edge.

As long as you keep trying to move the turner, the stick will stay stuck. When you get each of the sticks stuck against the turner, there will be nothing left to stop it, and it will turn and pull out the tooth.

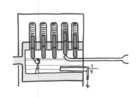

In some places, carrying these pieces of metal can get you in trouble even if you don't open anything with them.

That's sort of a strange law, since there's nothing wrong with using metal to turn a piece in a machine. Lots of people use these to learn how shape checker machines work.

But carrying these pieces of metal can make people worried for the same reason the machines are interesting—because they're not really machines. They're a way of telling people what you want to let them do. And that means these pieces of metal are also seen as messages—the idea that you don't care what other people want.

So it makes sense that people worry about them, even if you just want to learn about cool shape checkers.

And, of course, if you really need to open one of these machines, there are simpler ways to do it.

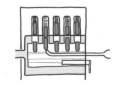
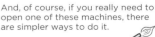

LIFTING ROOM

A lifting room is a box that carries people up and down in a building.

Today's cities wouldn't make sense without lifting rooms. If we had tall buildings without them, everyone would want to stay on their own floor, because going up or down would take a lot more work than going the same distance to the side. Tall buildings might have to join up with each other, and people would mostly move between them while staying on their own floors.

Most lifting rooms go straight up and down. A few go to the side while going up and down, to take people to the top of a hill.

There are also lifting rooms that *only* move side to side; those are called trains.

Lifting rooms are safe; there's almost no way they can fall. There are a lot of different parts that help lift them, and each part is made to stop the room—instead of letting it go—if something goes wrong.

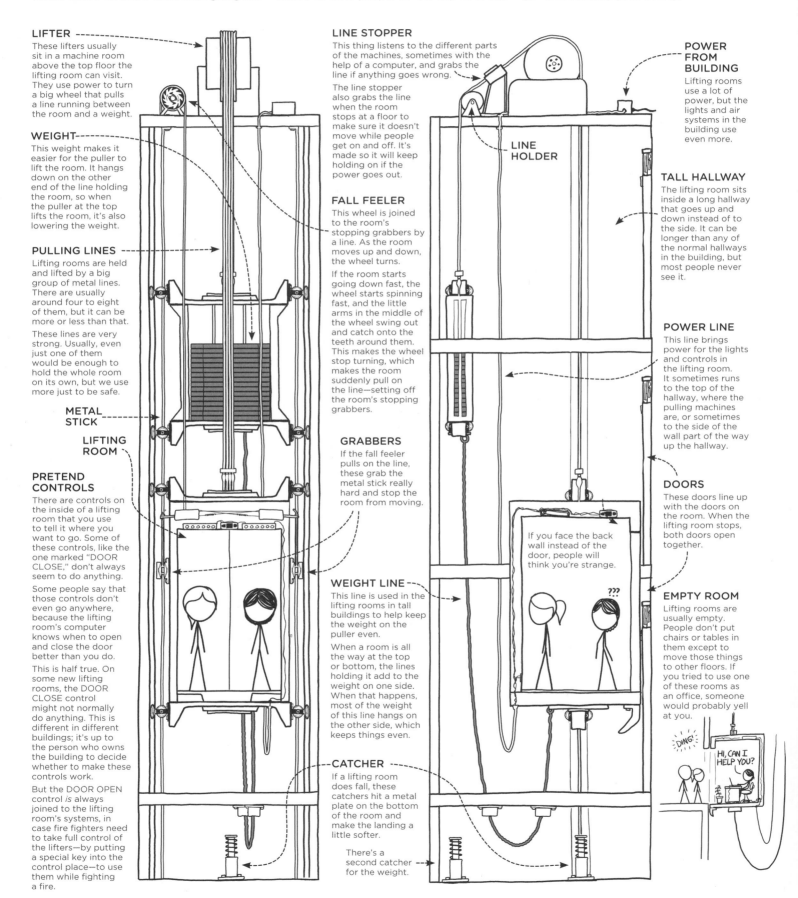

LIFTER

These lifters usually sit in a machine room above the top floor the lifting room can visit. They use power to turn a big wheel that pulls a line running between the room and a weight.

WEIGHT

This weight makes it easier for the puller to lift the room. It hangs down on the other end of the line holding the room, so when the puller at the top lifts the room, it's also lowering the weight.

PULLING LINES

Lifting rooms are held and lifted by a big group of metal lines. There are usually around four to eight of them, but it can be more or less than that.

These lines are very strong. Usually, even just one of them would be enough to hold the whole room on its own, but we use more just to be safe.

METAL STICK

LIFTING ROOM

PRETEND CONTROLS

There are controls on the inside of a lifting room that you use to tell it where you want to go. Some of these controls, like the one marked "DOOR CLOSE," don't always seem to do anything.

Some people say that those controls don't even go anywhere, because the lifting room's computer knows when to open and close the door better than you do.

This is half true. On some new lifting rooms, the DOOR CLOSE control might not normally do anything. This is different in different buildings; it's up to the person who owns the building to decide whether to make these controls work.

But the DOOR OPEN control *is* always joined to the lifting room's systems, in case fire fighters need to take full control of the lifters—by putting a special key into the control place—to use them while fighting a fire.

LINE STOPPER

This thing listens to the different parts of the machines, sometimes with the help of a computer, and grabs the line if anything goes wrong.

The line stopper also grabs the line when the room stops at a floor to make sure it doesn't move while people get on and off. It's made so it will keep holding on if the power goes out.

FALL FEELER

This wheel is joined to the room's stopping grabbers by a line. As the room moves up and down, the wheel turns.

If the room starts going down fast, the wheel starts spinning fast, and the little arms in the middle of the wheel swing out and catch onto the teeth around them. This makes the wheel stop turning, which makes the room suddenly pull on the line—setting off the room's stopping grabbers.

GRABBERS

If the fall feeler pulls on the line, these grab the metal stick really hard and stop the room from moving.

WEIGHT LINE

This line is used in the lifting rooms in tall buildings to help keep the weight on the puller even.

When a room is all the way at the top or bottom, the lines holding it add to the weight on one side. When that happens, most of the weight of this line hangs on the other side, which keeps things even.

CATCHER

If a lifting room does fall, these catchers hit a metal plate on the bottom of the room and make the landing a little softer.

There's a second catcher for the weight.

LINE HOLDER

If you face the back wall instead of the door, people will think you're strange.

???

POWER FROM BUILDING

Lifting rooms use a lot of power, but the lights and air systems in the building use even more.

TALL HALLWAY

The lifting room sits inside a long hallway that goes up and down instead of to the side. It can be longer than any of the normal hallways in the building, but most people never see it.

POWER LINE

This line brings power for the lights and controls in the lifting room. It sometimes runs to the top of the hallway, where the pulling machines are, or sometimes to the side of the wall part of the way up the hallway.

DOORS

These doors line up with the doors on the room. When the lifting room stops, both doors open together.

EMPTY ROOM

Lifting rooms are usually empty. People don't put chairs or tables in them except to move those things to other floors. If you tried to use one of these rooms as an office, someone would probably yell at you.

DING

HI, CAN I HELP YOU?

BOAT THAT GOES UNDER THE SEA

We've always had boats that go under the sea, but in the last few hundred years, we've learned to make ones that come back up.

At first, we used those boats to shoot at other boats, make holes in them, or stick things to them that blew up.

Later, we found a new use for these boats: keeping our city-burning machines hidden, safe, and ready to use if there's a war.

WORLD-ENDING BOAT

The boat shown here carries up to two dozen city-burning war machines.

People have added up the power used during the Second World War—all the machines that blew up, all the guns that fired, and all the cities that burned. It's a lot of fire power. Each of these boats carries several times that much.

SPECIAL SEA WORDS

Most of the time, if you call a really big boat a "boat," people who know a lot about boats will get mad at you. But boats that go under the sea are really called "boats."

HEAVY METAL POWER MACHINE

These boats are powered by heavy metal, just like some power buildings. This means they can stay hidden for a long time without running out of power.

Any time heavy metal is used for power, people worry about something going wrong. Of course, given what these boats are built for, people worry even more about the idea of one of them working *right*.

BREATHING STICK

This brings fresh air into the boat, but the boat can also make its own air by breaking water into the parts it's made of. This takes a lot of power, but the boat is powered by heavy metal, so it has enough power to do whatever it wants.

MIRROR LOOKERS

When the boat is hiding under the sea, it can come near the surface and use these sticks with mirrors in them to let the people inside see out of the water.

SOUND LOOKERS

Light can't go far under water, so these boats "see" with sound. The boat makes sound, which hits things and comes back. By listening carefully, the people in the boat can tell what's around them without seeing—just like those skin birds that catch flies in the dark.

SLEEPING ROOMS

The normal people on the boat sleep on either side of the city-burning machines.

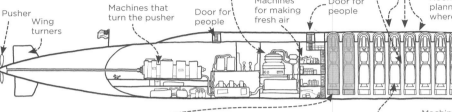

Pusher
Wing turners
Machines that turn the pusher
Door for people
Machines for making fresh air
Door for people
Doors for city-burning machines
Room for planning where to go
Offices
Kitchen
Room for making choices
Door for people
Radio

EMPTY ROOMS

A while ago, everyone decided the world didn't need so many city-burning machines. This country agreed to turn off four of the two dozen flying machine carriers in each boat, leaving only twenty.

MACHINES FOR BURNING CITIES

Each of these rooms has a flying carrier full of city-burning machines. While hiding under the water, the boats can shoot the machines into space. Any of these boats can send a machine anywhere in the world in under an hour.

Machine that makes power from fire water (in case there's a problem with the heavy metal)

Eating room
Flying carrier control room
Computers
Power boxes
Rooms that fill with water to make the boat go under the sea

MACHINES FOR SHOOTING BOATS

This boat can shoot these tiny machines under the water at other boats to make holes in them. They blow up, but don't use heavy metal.

Boats used to carry more guns and machines like this, but boats don't really fight each other anymore.

OTHER BOATS THAT GO UNDER THE SEA

These are some other boats, drawn to show how big they are next to the world-ending boat above.

WORLD WAR BOAT

This was used by one country in the Second World War. It was called an "Under-Sea Boat."

THE FIRST ATTACK BOAT

This boat was used over two hundred years ago to stick burning things to ships and blow them up.

SMALL ATTACK BOATS

These boats are big, but smaller than the ones that carry our city-burning machines. They carry machines that blow up houses, streets, and other boats, but not whole cities.

NEVER-USED BOAT

These were built over a hundred years ago, but kept hidden and never fought.

(That's not too strange; today's world-ending boats *also* hide and have never fought.)

DEEP GOER

Two people used this boat to visit the bottom of the deepest place in the sea.

BOAT FINDER

This was used to find a huge boat that had hit some ice long ago and fallen to the bottom of the sea.

MOVIE MAKER DEEP GOER

A man made a movie about the huge boat that hit some ice and broke, then used the money to buy this boat and take it to the deepest part of the sea. (He didn't go there to make a movie. He just likes the sea.)

Biggest animal
Biggest animal with teeth

BIG ANIMALS

These animals are smaller than our big fighting boats, but some of them can go much deeper.

HOW DEEP THEY GO

The sea is very deep. Most boats can't go very far down or their walls will break under the weight of the water. To go to the bottom of the sea in most places takes special boats.

PEOPLE

Most of the time, even with the help of metal air holders, people can't go deeper than a few hundred feet (at least, not if they want to come back up).

People *without* extra air sometimes also go that deep and come back up, but those people die a lot.

ANIMALS

The large air-breathing animal with teeth can go far, far down into the sea. They go there to eat animals with many arms.

Sometimes the air-breathing animals come back up covered in little cuts and holes, so the animals with arms must fight back, but no one has ever seen it happen.

FIGHTING BOATS

Most fighting boats can only go a few times farther down than than they are long. That's not very deep; when a boat is hiding under the sea, the water below it can be ten or even a hundred times deeper than the water above it.

But where they go is deep enough to hide and be safe, so they don't need to go any deeper.

THE DEEP SEA FLOOR

Only three people have ever been there—two in the deep goer and the movie maker.

BOX THAT CLEANS FOOD HOLDERS

This box is a machine that cleans plates and cups by throwing water at them. The water is full of cleaning stuff, which helps the water stick to the food and pull it off.

If you fill a cleaning box the wrong way, it may not clean well. After people see this happen a few times, they can get strong ideas about the right way to fill a plate cleaner. When people with different ideas about these machines start living together, this can even lead to fights.

Some ideas are clear to everyone—like that you should always point cups down, so they don't end up full of food water. There's more, but you don't need to fight over it! There's a book that goes with your plate cleaner, and it shows how you should fill it. (If you've lost the book, you can usually read it for free using a computer.)

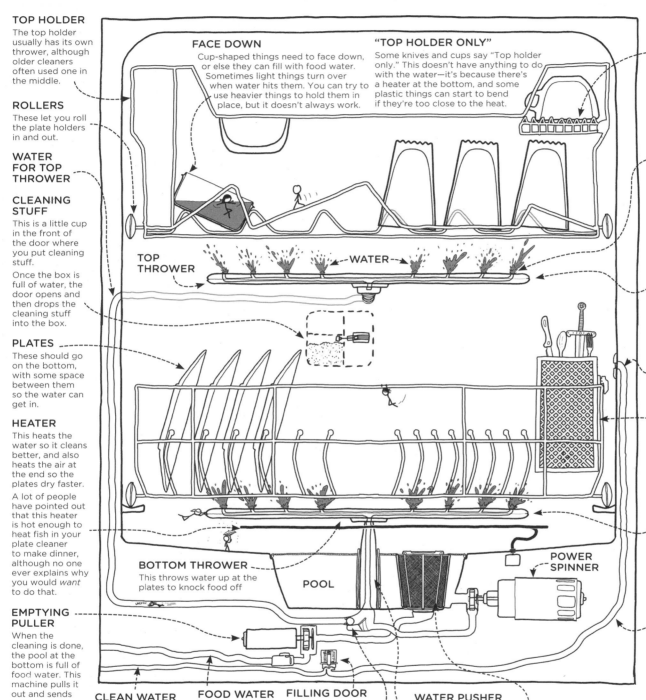

TOP HOLDER
The top holder usually has its own thrower, although older cleaners often used one in the middle.

ROLLERS
These let you roll the plate holders in and out.

WATER FOR TOP THROWER

CLEANING STUFF
This is a little cup in the front of the door where you put cleaning stuff.
Once the box is full of water, the door opens and then drops the cleaning stuff into the box.

PLATES
These should go on the bottom, with some space between them so the water can get in.

HEATER
This heats the water so it cleans better, and also heats the air at the end so the plates dry faster.
A lot of people have pointed out that this heater is not hot enough to heat fish in your plate cleaner to make dinner, although no one ever explains why you would want to do that.

EMPTYING PULLER
When the cleaning is done, the pool at the bottom is full of food water. This machine pulls it out and sends it away.

FACE DOWN
Cup-shaped things need to face down, or else they can fill with food water. Sometimes light things turn over when water hits them. You can try to use heavier things to hold them in place, but it doesn't always work.

"TOP HOLDER ONLY"
Some knives and cups say "Top holder only." This doesn't have anything to do with the water—it's because there's a heater at the bottom, and some plastic things can start to bend if they're too close to the heat.

TOP THROWER

← WATER →

BOTTOM THROWER
This throws water up at the plates to knock food off

POOL

POWER SPINNER

SMALL CUP HOLDER
This holder is for small cups, or knives, or anything that fits.
If you have something light that turns over a lot, you can put it under this holder.

TURNING
Throwers have holes that point a little to one side, so when the water goes out that way, it will push the thrower in the other direction. That's what makes them spin around.

DON'T BLOCK THIS
If you put a tall knife in the knife cup, it can block the top thrower from turning. If it can't turn, the water won't reach all your plates.

FILLER HOLE
Clean water comes in here.

KNIFE CUP
This cup holds knives and other pointy things. You should *always* point sharp knives down, so if you trip and fall on them, they won't hurt you.

DON'T BLOCK THIS, EITHER
If you put a knife here, instead of the knife cup, it can fall through and stop the lower arm from spinning.

FILLER LINE
This line brings in hot water from the house and pours it into the box at the start.

CLEAN WATER COMING IN

FOOD WATER GOING OUT

FILLING DOOR
This opens when it's time to fill the pool.

WATER PUSHER
Water gathers in a pool at the bottom of the box. This pusher takes water from the pool and pushes it up to the throwers.

To top thrower   To bottom thrower   Water pusher

IN DIFFERENT CLEANING BOXES THIS WORKS IN DIFFERENT WAYS, BUT I LIKE THIS KIND BECAUSE IT'S A COOL IDEA I WOULD NEVER HAVE THOUGHT OF.

THROWER CONTROL BALL
This ball controls which thrower to send water to.

At the start, the ball sits at the bottom of a little slide, blocking the path to the top thrower. When the water pusher runs, the water goes to the bottom thrower.

The ball doesn't block *all* the water, and a little water starts to get past the ball and fill up the path to the top thrower.

When it's time to use the top thrower, the water pusher stops for just a moment. The water behind the ball falls back down, pushing the ball up the slide.

Just as the ball reaches the top of the slide, the pusher turns back on. The force of the water holds the ball against the hole as it runs to the top thrower.

When the pusher is done, it stops for a moment, and the ball rolls back down the slide to where it started.

FOOD CATCHER
This catches pieces of food so they don't go into the water pusher, since they could get sent through the machine and get stuck.

The catcher has a hole in the bottom so food can go into the emptying puller instead. If your cleaning box stops working, you may need to clean this.

A PROBLEM THAT CAN HAPPEN
The line that carries food water out of the pool joins with your other bad-water carriers before it leaves the house.

It needs to go up high under the table top. If it runs straight into the hole under the other water catcher, then if the catcher gets full for some reason, bad water can run back into the cleaning box.

Cleaning box

Hand wash area

Bad water

BIG FLAT ROCKS WE LIVE ON

The surface of the Earth is made up of big flat rocks moving around. The rocks under land areas are usually thick, slow-moving, and last for a long time, and the ones under seas are thin, heavy, and move fast. (Fast for a rock, that is. They move about as fast as the things on the ends of your fingers grow.) When a sea rock hits a land rock, the sea rock is usually pushed under it, down into the Earth. Areas where this happens often have deep seas right near land, lines of mountains, shaking ground, and big waves.

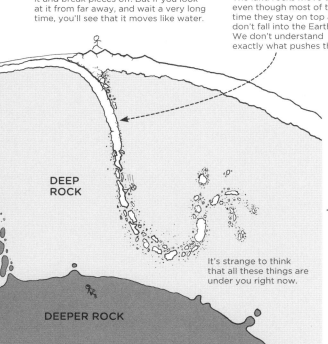

HOT ROCK MOUNTAIN
The rocks that get pushed into the Earth get hot and watery, and some of them come up through holes in the rock above them. They come out of those holes and cool down and turn into mountains.

ROCK MOUNTAIN
Not all the mountains in this kind of place are made from hot rock. When a sea plate goes under a land plate, it can make mountains by pushing up on the land plate.

If two land plates hit, it can make very big mountains. The biggest mountains on Earth right now were made this way.

DEEP PART
The sea floor is deeper here because the sea rocks are getting pushed down as they run into the land rocks.

LOW AREA
These low areas between mountains sometimes have water in them, and the ground there is usually good for growing things, so people like to live there. Sometimes hot rocks come out of the mountains and cover everyone's houses. But that doesn't happen very often, so people try not to worry about it too much.

Small Soft, the company that makes Windows®, is in a city like this.

SEA FLOOR

LAND FLOOR

LAND ROCKS
These are like big rock boats that drive around on top of the hotter rocks under them.

SEA ROCKS
Sea rocks are heavy. They slide along like a moving road, and they move fast! Not as fast as a person, but faster than most kinds of land.

When sea rocks hit land rocks, the sea rocks usually get pushed back down under the land rocks, down into the Earth, where they break down. Because most sea rocks run into land and disappear after a while, most parts of the sea floor aren't as old as the land floor.

WATER-CARRYING ROCKS
Sea water gets carried inside the Earth here. The water changes the rocks in a way that helps the rock go back up through the rocks above it and come out of holes in the ground.

DEEP ROCK
This part of the world can be hard to understand. Sometimes people talk about it like it's watery, but sometimes they talk about it like it's hard.

The real truth is that it's very hard. If you touched a piece of it, it would feel very hard. (You shouldn't touch it, though, because it would also set your hand on fire.) It's harder than the hardest metal, glass, or even stones in a marriage ring. That makes it sound a lot like rock, not water.

But in some ways, it also acts like water. It's sort of like the big rivers of ice that slide slowly down from mountains. The ice is hard up close, and you can walk on it and break pieces off. But if you look at it from far away, and wait a very long time, you'll see that it moves like water.

The reason the sea floor moves is that these rocks are heavier than the deep rock below them, and their weight pulls on the sea floor as they fall down into the Earth.

The land rocks move too, even though most of the time they stay on top and don't fall into the Earth. We don't understand exactly what pushes them.

DEEP ROCK

It's strange to think that all these things are under you right now.

DEEPER ROCK

WHEN THE EARTH SHAKES, SOMETIMES THERE ARE BIG WAVES. THIS IS THE KIND OF SHAKING THAT MAKES THE BIGGEST WAVES:

There's a place in my country on the edge of the sea. (They once made a game for kids about trying to get to this place. You had to cross rivers and shoot animals for food and sometimes people in your family died. It was supposed to teach you about the past, but I just played the shooting part and never learned very much.)

Right by the water, there's something very strange: dead trees in the sea. There are lots of dead trees in the sea. But what's strange about these trees is that they're not lying down. They're sticking up from the sea floor, like they grew there. That shouldn't be possible, because those trees can't grow in sea water. The sea rises and falls, but the trees only died 300 years ago, and the sea hasn't risen enough to explain how these trees grew there.

The answer is that the sea didn't rise. The land fell.

On the other side of the sea, 300 years ago, there was a big wave. People who saw it wrote about it. They also wrote that they didn't feel the ground shake before the wave.

The reason they didn't feel the ground shake is that the shaking didn't happen near them. It happened far away, across the sea, in the place from the kids' game. And by the edge of the water, the ground went down a little bit, and the sea came in and covered the trees.

WHERE DO ROCKS GO AFTER THEY DIE?

We used to think that when rocks fell into the Earth, they broke up right away from the heat. And even if they stayed together for a little while, it didn't really matter, since they were hidden away forever. That part of our history was gone.

But it turns out they're not quite gone. When the world shakes, we can listen to the sound go around and through the world. By listening very carefully, we can hear the sound hitting things inside the Earth, and learn what it's like in there.

By listening to the Earth, we've learned that the rocks don't all turn to water right away. We can keep track of them, even when they're out of reach of our eyes, as they fall down, down, down into the Earth.

I think that's really cool.

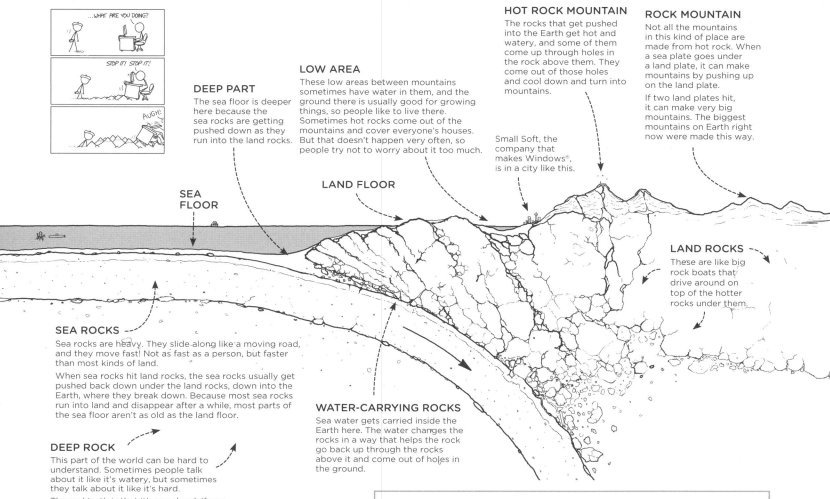

CLOUD MAPS

The air changes every day. Every day, clouds move around, rain comes and goes, and winds change. And every day, people all over the world try to figure out what the air is doing and where the rain will go next.

To make maps of the sky, we use space boats looking at clouds from above, radio waves looking at clouds from the side, and people all over the world looking at clouds from below.

HIGHS AND LOWS

These lines show how hard the air is pressing down on different areas of the map—which is sort of a strange idea, but important for understanding rain and wind.

These maps are a lot like maps used to show the shape of mountains. The lines join areas where the air is pressing down with the same weight, and the middles of circles are areas where air is especially heavy or light. They're marked "Heavy" and "Light" (or "High" and "Low") to help you know which is which.

LOWS (RAIN MAKERS)

Areas with lighter air over them are called "lows." Air moves across the ground toward those areas, and—just like water moving toward a hole in the bottom of a pool—it goes faster and starts moving in a circle.

Air usually rises up in these "light" areas, which makes rain. As the air rises, the water in the air cools down and turns into little drops, just like water on the side of a glass with a cold drink in it.

COLD AIR

This area will be cold and clear.

This area will have heavy rain (or snow, if it's cold enough.)

This area will have strong, cold winds and heavy rain.

This area will have light wind and light rain.

COOL AIR

COOL AIR

This area will be cool.

HIGH

HIGHS (CLEAR AREAS)

In a "heavy" (or "high") area, air is pressing down hard, which keeps wet air from rising and keeps clouds and rain from forming. These areas usually have clear skies and not very much wind.

This area will be clear and warm for now.

The dark areas on this map show where it will rain.

WARM AIR

This area may see flashes of light in the sky and winds strong enough to blow away a house.

GREAT CIRCLE STORMS

These storms are a kind of "low" powered by the heat carried by sea water as it turns to air and rises from the surface when warmed by the Sun. They have very strong winds in a circle near the center, but right *in* the center it's calm—and can even be clear. People call this clear area the "eye" of the storm.

When these storms come in from the sea, they bring the sea with them. Their winds push water ahead of them, and it can make the sea come up onto the land and cover whole cities. They can also make so much rain that rivers rise and wash away people, cars, and houses.

Thanks to computers, we've gotten a lot better at guessing where circle storms are going to go, which helps us to tell people to get out of the way.

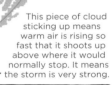

This piece of cloud sticking up means warm air is rising so fast that it shoots up above where it would normally stop. It means the storm is very strong.

Around here, the air stops getting colder as you go higher, so warm air stops rising.

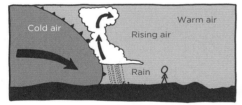

COLD AIR COMING IN

This line shows where cold air is coming in. This can mean there will be wind, and then flashes of light, sounds from the clouds, and very, very heavy rain, but it doesn't last long.

WARM AIR COMING IN

This line means warm air will be moving into an area. This can mean there will be clouds ahead of the warm air, sometimes a few days before it gets there, and rain as it moves in.

VERY BIG SUMMER STORMS

Sometimes, on hot days, air heated by the sun rises up very fast, then cools and pours down rain. These storms can make spinning wind that blows away houses.

THINGS YOU SEE ON RADIO MAPS AND WHAT THEY MEAN

Sky-watching stations point radio waves at clouds. If there are big drops of water in the clouds, the radio waves hit them and come back. By pointing the radio waves in different directions, the people in the stations can make a map of all the rain and snow in clouds around them.

Here's how to understand some of the shapes you see on those maps:

RAIN

Big shapes like this mean rain. It will probably last a while and be light sometimes and heavy other times.

SOUND STORM

This shape means a storm is coming, which may bring light, sound, and strong wind.

WIND STORM

This shape means a storm with lights and sound is coming, and the wind ahead of it might be even stronger than normal.

SPINNING WIND

This shape, like a bent finger, means a spinning cloud is touching the ground, and may be tearing up trees and houses.

Sometimes, if you look at the shapes made by the radio, you can see the stuff the storm has picked up. It looks like a small ball in the middle of the bent finger shape.

SKIN BIRDS

This circle shape isn't rain—it's hundreds and hundreds of little skin birds all flying out of a big hole to eat flies when the Sun sets.

Sometimes other animals, like normal birds or flies, show up on these maps too.

TREES

When there's no rain to see, sometimes the map shows little lines of noise from the radio waves hitting the tops of trees and houses.

GROUND

This shape happens when the radio waves hit clouds, then a pool of water on the ground, then come back. That makes them take longer, so it looks like there's rain far away.

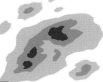

TREE

LEAVES
Trees make power from the Sun's light using leaves. The green stuff in leaves eats light (and the kind of air we breathe out) and turns it into power (and the kind of air we breathe in).

GRAY TREE-JUMPER
These little animals sleep in big round houses made of sticks and leaves high up in the branches.

GROWING UP
Trees grow taller only by making the ends of their branches longer. The spot where a branch joins the main part of the tree is never lifted higher.

POINTY CAT
This animal walks around slowly, climbing trees and eating leaves and sticks. It's covered in sharp points that can stick in your skin, so most animals don't bother it.

BIRD HOLES
Some birds make holes, but a lot of them just use holes other birds make.

TALL AND WIDE TREES
The same kind of tree can grow tall or wide. If there are other trees around, they'll grow mostly up, each one trying to get above the others to reach the Sun's light. If a tree is growing alone in a field, it will spread branches out to the sides so it can catch more light.

QUIET NIGHT CATCHER
These birds fly very quietly and have big eyes to catch animals on the ground in the dark.

People think of them as knowing a lot of things, although that may just be because they're quiet and have big eyes.

TREE-EATING FLOWERS
This flower makes holes in trees and steals food and water from inside them. If the flowers get big, they can kill the branches they're growing on, or even kill the whole tree.

When people stand under this flower at a party, other people tell them to kiss.

LOUD JUMPERS
These two kinds of tiny animals make loud noises and are known for jumping. One has bones.

DRINK HOLES
These were made by a head-hitting bird looking for tree blood to drink.

FIELD TURNING INTO FOREST
When people cut down a forest, sometimes they leave a few trees—to make a cool shadow area, or because the tree looks nice—and those trees will grow out into the new space.

If the forest grows back, the new trees—fighting with each other as they grow—will be tall and thin.

If you find a forest of tall thin trees with one wide tree with low branches in the middle, it might mean the forest you're in was someone's field a hundred years ago.

SKIN BURNER
These leaves have stuff on them that makes your skin turn red. It gives you a really bad feeling, like you need to rub your skin with something sharp, but doing that only makes it worse.

This leaf-flower grows in long lines across the ground or up trees. Sometimes it grows into the air like a small tree of its own. Like many things, its leaves come in groups of three.

STORM BURN
When flashes of power from storms hit a tree, they can burn a line in the wood.

BROKEN BRANCH HOLE
When a tree gets hurt, like if a branch breaks off, the place where it got hurt grows differently, just like when skin gets cut. Sometimes animals get in through these spots and make the hole bigger.

HEAD-HITTING BIRD
This kind of bird hits trees with its head, making holes in the wood with its sharp mouth. They make holes to find things to eat, and some also make holes to live in.

BIRD HOUSE

SKIN
The outer skin of trees is where growing happens and where they carry food up and down. Cutting off a ring of skin all the way around a tree will kill it.

Trees grow by adding new layers, and grow differently in different parts of the year. If you cut open a tree, you can see old layers, and count them to tell how many years old the tree is.

OLD METAL
When people use metal to stick signs to trees, sometimes the tree grows around the metal and eats it up.

Then, many years later, if someone needs to cut down the tree, their saw can hit the metal and send tiny sharp pieces flying everywhere.

ANIMAL HILL
This is the dirt the walking flies took out of the ground while making their holes.

DOOR

DIRT BRANCHES
Trees grow branches down into the ground, like the ones in the air. The air branches get light from the Sun, while the ground branches get water and food from the dirt. They spread way out—often farther than the air branches—but usually not very deep.

FIRE HOLE
These holes are from fires long ago. The leaves and sticks on the ground burned, and the wind blew the fire against this side of the tree. The burned spot grows in a different way and can sometimes turn into a large hole.

TREE-FOOD STEALER
Instead of growing dirt branches of their own, these flowers grow onto the dirt branches of other trees and steal food from them.

Some of these little flowers don't even have green leaves and can't make their own food from light.

TINY DOG

LITTLE HOLE-MAKERS

LONG-EAR JUMPERS

BIG HOLE-MAKERS

WALKING FLIES
These tiny animals live in big groups and make holes. Most of them don't have babies; each family has one mother who makes all the new animals for the house.

They usually don't fly, and they're not much like house flies. They're in the same group with the kinds of flies whose back end has a sharp point that can hurt you.

LONG BITERS WITHOUT ARMS OR LEGS (SLEEPING)
These long thin cold-blooded animals don't usually hang out together, and sometimes eat each other.

During the winter, though, lots of different kinds come together and sleep all wrapped up together in big holes under the ground where it's warmer.

LONG-HOLE MAKERS

DIRT-BRANCH LIFE
Most trees and flowers have life growing on their dirt branches. This life helps them talk to the other trees and flowers around them. They can even use this life to share food or attack each other.

If something tries to eat one tree, it can tell other trees through messages carried by this ground life, and the other trees can start making bad water and other things to make themselves harder to eat.

MACHINE FOR BURNING CITIES

At the end of the biggest war in history, less than a hundred years before this book was written, some people figured out how to make a small piece of heavy metal heat up like the Sun. They could make the metal get so hot that it would blow up with enough light and fire to burn a whole city, and send up clouds of dust that made people sick. Two of the machines were used in that war, and each one burned a city and killed many, many people.

After the war, we learned to make the fire from the machines even bigger and hotter, and built flying carriers that could send them anywhere in the world in just a few minutes. There was no way to stop these machines, so many countries built them, and hid them under the ground, so that no one could attack them without getting attacked back.

Everyone worried that a new war would start at any minute. We spent many years like that, each side waiting for the other to attack and start the war that would end the world.

We're less afraid now, and most people don't think that war will happen. But we still have the machines.

THE MACHINE

The first machines had one part that blew up, but a few years later, we learned to make the fire a lot bigger by putting two parts together.

The top part uses a normal fire to start the run-away fire in the special metal. Then the bottom part uses that special fire to start an even bigger run-away fire in a light air or metal. That second fire is the kind that powers the Sun.

The light metal's run-away fire can let out even more power than the heavy metal kind, but it takes so much heat and force to get it started that we can only do it with the help of a run-away fire in a heavy metal.

THE FIRST RUN-AWAY FIRE

Everything is made of tiny pieces. Around the start of the Second World War, we learned that the pieces of a few special heavy metals could be made to break in half. We also learned that when they break, they let out a little flash of heat and some tiny fast pieces.

THE PIECES EVERYTHING IS MADE OF

Cloud part

Heavy center part

The clouds of little things that fly around the heavy center part aren't important for the run-away fire; we can ignore them.

RUN-AWAY FIRE

When the heavy center of one of the metal pieces breaks in half, it lets out heat and some pieces. If those pieces hit another center, the same thing happens, letting out more heat and pieces. Soon, the whole piece of metal can become a run-away fire.

ENOUGH METAL

If the piece of metal is too small, the little pieces from the broken center can fly out without hitting any other centers.

For a run-away fire to start, there needs to be enough metal to make sure the flying pieces hit other centers instead of flying out.

(Don't do this.)

HOW MUCH IS "ENOUGH"?

The size of a piece of metal it takes to make a run-away fire is different for different metals and different shapes, but it can be a piece small enough for a person to pick up.

Even if a piece isn't big enough, pushing it into a small space can make it blow up, because when the centers are closer together, there's less room between them for the fire to get out.

FIRE PLASTIC

This is the normal kind of stuff people use to blow things up.

FIRST PART

STARTER

TINY FIRE MAKER

HOLE

This is for adding some special air, right before the machine goes off, to help the run-away fire start.

NORMAL METAL

This helps hold the special heavy metal together as the run-away fire starts.

HEAVY METAL

This is where the first run-away fire happens.

IN-BETWEEN STUFF

We don't know what this is made of; the machine makers have kept that hidden. When the light fills the inside of the box, it gets bigger and pushes the second part together.

SECOND PART

WALL

This helps hold in the light from the first part so it can push on the second part.

LIGHT METAL OR AIR

This stuff can also burn in a run-away fire, but it has to be pushed together really hard first.

MORE HEAVY METAL

When the light metal or air is pushed together, it also starts another run-away fire here. These run-away fires help make each other stronger.

THE SECOND RUN-AWAY FIRE

Here's how the first run-away fire sets off the second one.

First, a message goes down a line and starts some tiny fires.

The tiny fires set off the fire plastic, which starts to blow up.

The blowing-up plastic pushes the heavy metal together.

When the metal gets small enough, the run-away fire starts.

As it burns, the metal puts out a bright light—brighter than anything but a dying star.

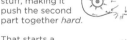

The light heats up the in-between stuff, making it push the second part together *hard*.

That starts a run-away fire in the *light* metal.

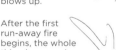

This fire makes the other fire even worse, and the whole thing blows up.

After the first run-away fire begins, the whole thing happens in the time it takes light to go a few hundred feet.

By adding more and more steps like this, we found we could make the fires as big as we wanted, and at first we built the machines larger and larger.

But then we stopped making the machines larger, and started making them smaller, instead. We didn't stop because we didn't want to burn larger cities. We just realized you could burn a city more easily with a few small machines than one big one. Soon, we had enough small machines to burn as many cities as we wanted.

We stopped making the machines larger because the ones we had were big enough to burn everything. There was nothing larger to burn.

HOW WE SEND THEM

The first city-burning war machines were dropped from sky boats. Later, we learned to put them on fast up goers instead.

City-burning machine carriers work a lot like the up goers that fly people up to space.

In fact, some of those people carriers are just city-burner carriers without the city-burning part at the end.

GOING TO SPACE (BUT NOT FOR LONG)

The city-burning machine carrier flies up to space. Like most up goers, the machine carrier drops parts once they're used up, so it can go faster and faster.

It goes almost fast enough to stay in space and go around the Earth.

Almost, but not quite.

24

WATER ROOM

This is one of the best things humans have ever built.

Over the past few hundred years, we've learned a lot about why people get sick. We've learned how the things travel that make us sick, and we've learned ways to stop them.

We've learned that often, when we get sick, it's because a kind of life has gotten into our bodies and is trying to grow in us. Our body can often fight this off, but while we're fighting it, it uses the things that come out of our body—often, by helping more of them come out—to spread itself to other people.

By learning how to bring water into our buildings, and how to make it carry things away from our bodies without touching others, we've found a way to fight back against the things that have killed so many of us.

This is an important room!

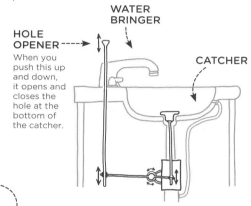

HOLE OPENER
When you push this up and down, it opens and closes the hole at the bottom of the catcher.

WATER BRINGER

CATCHER

AIR HOLE TO ROOF
There's usually some air in the smelly water that's going out of the room. This carrier lets that air rise up and out a hole in the roof, instead of coming back up through the hole in your room and making everything smell bad.

SMELLY AIR

WALL SOUNDS
Sometimes, when you turn off the water in an old house, you hear a loud sound from the wall, like a big rock hitting something. That's the sound of water hitting a stopper.

When you turn on the water bringer in a house, there's a long train of water all moving together toward the hole where it's coming out. When you turn it back off, all that water has to stop.

Water moves around a lot, but it can't really get smaller. When the front part of the water train hits the stopper, since it can't get smaller, there's no place for it to go. It all has to stop right away. The force of the water train all stopping at once hits the metal around it really hard, making a loud noise.

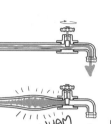

OLD HOUSE:

NEW HOUSE:

WHAM

HOW WE FIXED THEM
Newer houses fix this by adding a piece to the water bringer. The piece is a dead-end area, above the water, which is full of air. When the water stops, it can go up into the dead end. The soft air in the dead end acts like a spring, slowing the water down gently so there's no loud noise.

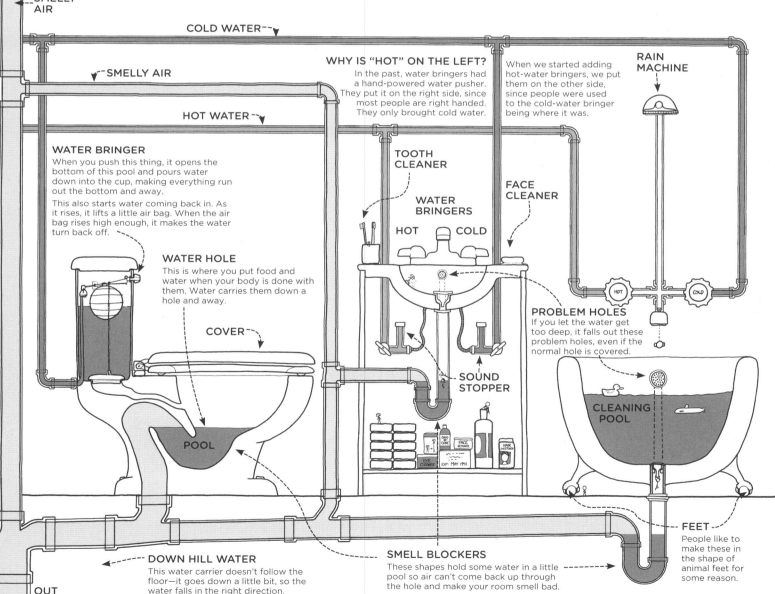

COLD WATER

SMELLY AIR

HOT WATER

WHY IS "HOT" ON THE LEFT?
In the past, water bringers had a hand-powered water pusher. They put it on the right side, since most people are right handed. They only brought cold water.

When we started adding hot-water bringers, we put them on the other side, since people were used to the cold-water bringer being where it was.

RAIN MACHINE

WATER BRINGER
When you push this thing, it opens the bottom of this pool and pours water down into the cup, making everything run out the bottom and away.

This also starts water coming back in. As it rises, it lifts a little air bag. When the air bag rises high enough, it makes the water turn back off.

TOOTH CLEANER

WATER BRINGERS

HOT   COLD

FACE CLEANER

WATER HOLE
This is where you put food and water when your body is done with them. Water carries them down a hole and away.

COVER

POOL

SOUND STOPPER

PROBLEM HOLES
If you let the water get too deep, it falls out these problem holes, even if the normal hole is covered.

HOT   COLD

CLEANING POOL

DOWN HILL WATER
This water carrier doesn't follow the floor—it goes down a little bit, so the water falls in the right direction.

OUT

SMELL BLOCKERS
These shapes hold some water in a little pool so air can't come back up through the hole and make your room smell bad.

FEET
People like to make these in the shape of animal feet for some reason.

COMPUTER BUILDING

When you use a computer to listen to songs or watch movies, sometimes they're on your computer, but often they're "in the cloud."

"The cloud" is just a lot of buildings owned by big companies. They're full of rows of computers, information boxes, and lots of confusing colored lines everywhere joining everything together, carrying information and power in and out of the computers. When you use things like Short Bird Talk and Face Book, your computer is talking to computers in buildings like this.

Some computer buildings were built by companies to hold all their own computers. Some very big companies do this. Other computer buildings sell space to people who have computers but need somewhere to put them. Some let you use their computers if you pay. Many have computers, and if you pay them, they use their computers to do things for you. But the computers in all these buildings usually look about the same.

FIRE-STOPPING AIR

If something catches fire in a computer building, the building's systems will often open up boxes full of heavy air. Fire needs stuff that's in our air so it can burn. If you put another kind of air in its place, the fire stops.

(The special air fires need to burn is the same kind of air we need to breathe, so if you're in the room when they send in the other air, you can die. But at least you won't be on fire.)

WAIT CONTROL

If there's a fire, and the building starts to fill itself with fire-stopping air, press this control. It tells the building, "Wait, don't change the air yet! I'm still here!"

COOLERS

Computers make heat. One of the hardest things about running a computer building is keeping it cool. A lot of the power these buildings use goes for running blowers in every room, pushers to move cooling water up to the roof and back, and running the big coolers on the roof that cool the cooling water down.

COLD AND WARM HALLS

Floors are laid out in rows, with cold halls and warm halls between them. The cold halls are on the side where computers pull air in, and the warm halls are on the side where the air is blown out of the computers. That way, computers aren't blowing hot air into other computers on the side where they're pulling air in.

SPECIAL ROOMS

If you have your own computers and computer holders, you can pay for a room to yourself. You can bring any computers you want and join them up with the building's systems.

MEET-ME ROOM (That's the real name.)

Sometimes, different companies have computers in the same computer building, and they want to send things to each other.

Normally, they would have to send stuff out of the building into the world, and pay information-carrying companies to bring it to the other company's computers—even if those computers were in the same building where it started. Some computer buildings have a room where different companies can all join up their computers and share stuff with each other without having to go outside or pay other companies to carry their messages.

POWER SENDER BOX

These boxes send power to each row of computers on the floor.

BLOWER SOUND

Computer buildings are loud. Most of the sound comes from all the blowers keeping parts cool.

PART FIXER BAG

This was left here by the people who own the room.

LOCKS

Computer buildings usually have at least two doors with machines that only let certain people in.

People who own computer buildings worry about this a lot, because if someone stole stuff from them, no one would want to leave their computers there anymore.

PERSON CATCHER

People have to close the outside door before opening the inside one. This keeps people from walking in behind you while the door is open.

NORMAL AIR

INFORMATION LINES

The lines to and from the computer rows sometimes run along the ceiling and sometimes along the floor.

COOLING WATER CARRIER

ROOM COOLER

There's a box like this on each floor. It feels the air in the room, and if it's too hot, it uses special cold water from the coolers on the roof to cool it down.

FIRE-STOPPING AIR

WARM HALLS

COLD HALLS

BEEP BOOP I AM FACE BOOK

OOPS

LIFTING ROOM

ROLLING PROBLEM TABLE

ROWS OF COMPUTER HOLDERS

BUILDING OFFICE

GUARD

OUTSIDE LINES

These join the building with the world's computer and phone systems. The lines are made of glass, instead of metal, which lets them carry more information.

FIRE WATER

POWER BOXES

KEEPING THE POWER ON

The people who run computer buildings worry a lot about the power going out. They usually have lots of power boxes to keep themselves running for a short time if the power dies, and some machines for burning fire water to make power if the power stays off for a long time.

POWER WATCHER

This machine decides where to get the power to send up to the computer floors. If the outside power dies, the machine changes over to using the power boxes without letting anything turn off.

POWER CHANGER

Computer buildings use a lot of power, so the kind of power they get from the power company isn't like the power to a house—that kind is too hard to send a long way. Instead, they get the kind of power that runs through those long lines that you sometimes see hanging from tall metal things high above the trees out in the country.

These boxes change that power into the normal kind of power computers need. They make things a lot easier, except when they blow up. But most days they don't blow up.

FINGER CHECKER

These machines have pictures of the lines on the fingers of everyone who's allowed in. When you touch them, they check the lines on your fingers and only open if the lines make you look like one of the people who's allowed in.

COMPUTER

Computer buildings use a special kind of computer that fits well in a holder. It's about the size of the back of a chair.

POWER SENDER
This takes power from the outside and sends it to different parts of the computer.

BLOWERS
These blow air through the computer to keep it cool. The air always blows from the front to the back, into the warm hallway.

MEMORY STICKS
These hold things the computer is thinking about, like information it's sending or pictures it's looking at. If the computer turns off, this goes away.

BACK
If you want to add a line into the computer, you usually run it in through a hole here.

CARD-HOLDER
These are spaces where you can put in a few more smaller add-on computers. These might do things like talk to other computers faster than normal, or do special work with numbers.

THINKING BOXES
These are the main centers where the computer follows steps and moves numbers around.

GRABBERS
These join the computer to the computer holder.

INFORMATION BOXES
These boxes remember things even when the computer turns off.

FRONT
This part usually has some lights to tell you how the computer is doing, and a little sticker with a name or picture that tells you what company made the machine. Since you're usually seeing it because you're fixing a problem, the sticker just tells you who you should be angry at.

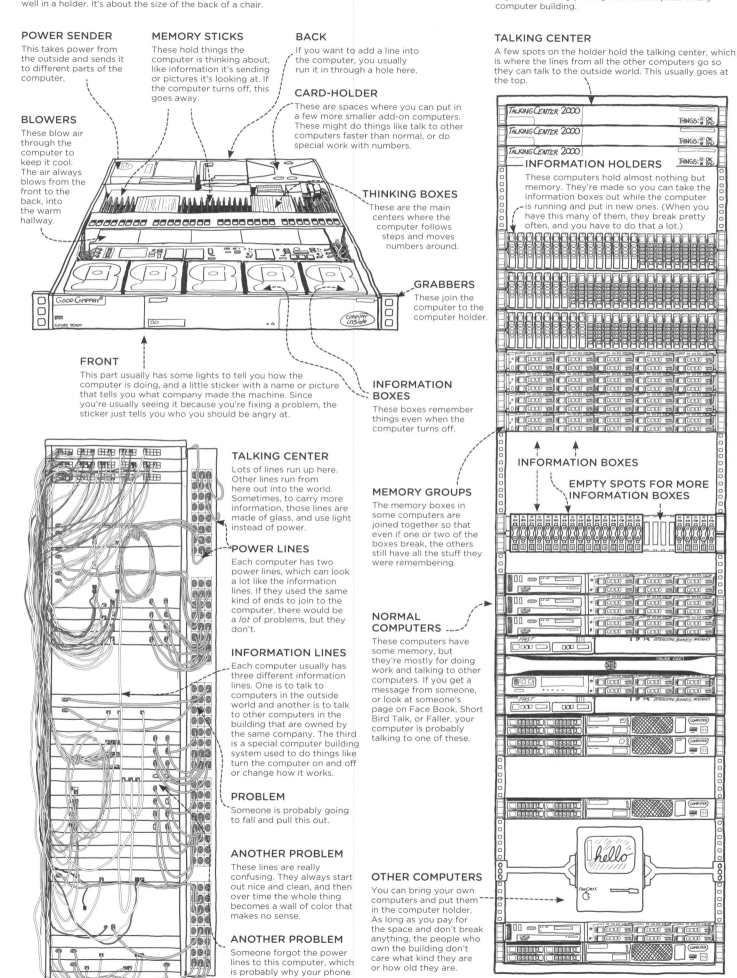

TALKING CENTER
Lots of lines run up here. Other lines run from here out into the world. Sometimes, to carry more information, those lines are made of glass, and use light instead of power.

POWER LINES
Each computer has two power lines, which can look a lot like the information lines. If they used the same kind of ends to join to the computer, there would be a *lot* of problems, but they don't.

INFORMATION LINES
Each computer usually has three different information lines. One is to talk to computers in the outside world and another is to talk to other computers in the building that are owned by the same company. The third is a special computer building system used to do things like turn the computer on and off or change how it works.

PROBLEM
Someone is probably going to fall and pull this out.

ANOTHER PROBLEM
These lines are really confusing. They always start out nice and clean, and then over time the whole thing becomes a wall of color that makes no sense.

ANOTHER PROBLEM
Someone forgot the power lines to this computer, which is probably why your phone has no service right now.

MEMORY GROUPS
The memory boxes in some computers are joined together so that even if one or two of the boxes break, the others still have all the stuff they were remembering.

NORMAL COMPUTERS
These computers have some memory, but they're mostly for doing work and talking to other computers. If you get a message from someone, or look at someone's page on Face Book, Short Bird Talk, or Faller, your computer is probably talking to one of these.

OTHER COMPUTERS
You can bring your own computers and put them in the computer holder. As long as you pay for the space and don't break anything, the people who own the building don't care what kind they are or how old they are.

HOLDER

Computer buildings keep all their computers in these holders. A holder can hold all kinds of computer parts; anyone can easily put any kind of computer in any computer building.

TALKING CENTER
A few spots on the holder hold the talking center, which is where the lines from all the other computers go so they can talk to the outside world. This usually goes at the top.

INFORMATION HOLDERS
These computers hold almost nothing but memory. They're made so you can take the information boxes out while the computer is running and put in new ones. (When you have this many of them, they break pretty often, and you have to do that a lot.)

INFORMATION BOXES

EMPTY SPOTS FOR MORE INFORMATION BOXES

US SPACE TEAM'S UP GOER FIVE

This is the only space boat that's landed people on another world. People landed on the Moon with it six times, all about half a hundred years before this book was written.

After those visits to the Moon, we stopped using this space boat to go to other worlds. The US Space Team used the boat, one last time, to send up their first space house.

After people visited the house a few times, it fell back down. Pieces of it landed in a small town. The town told the US Space Team to pay a fine for dropping stuff on the ground.

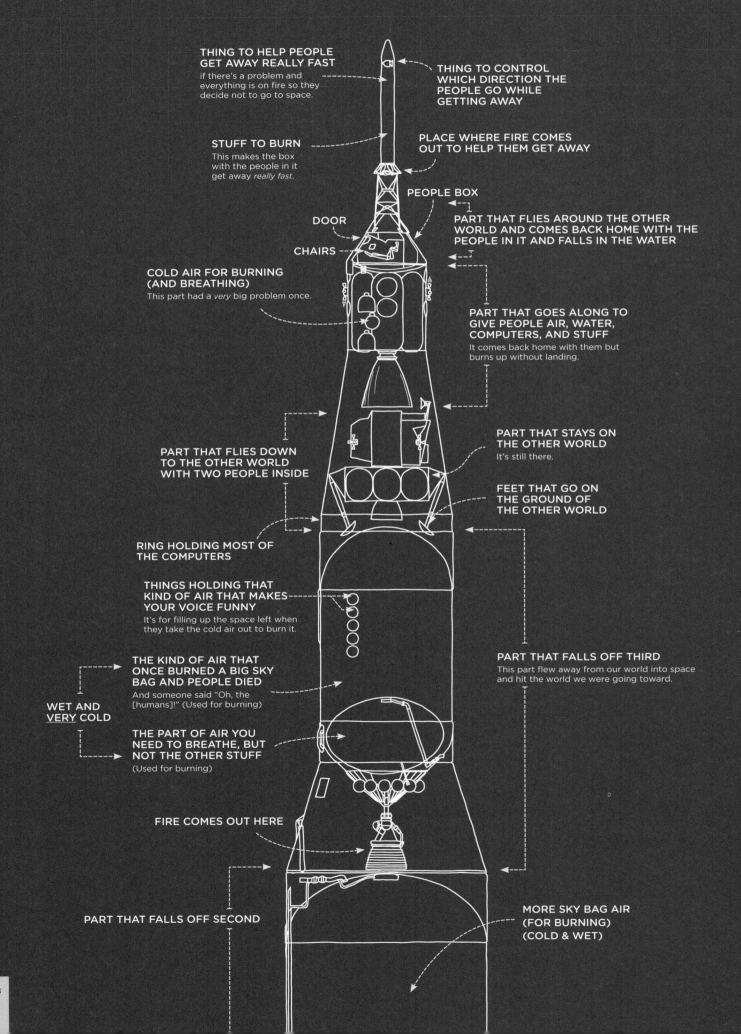

THING TO HELP PEOPLE GET AWAY REALLY FAST
if there's a problem and everything is on fire so they decide not to go to space.

THING TO CONTROL WHICH DIRECTION THE PEOPLE GO WHILE GETTING AWAY

STUFF TO BURN
This makes the box with the people in it get away *really fast*.

PLACE WHERE FIRE COMES OUT TO HELP THEM GET AWAY

PEOPLE BOX

DOOR

PART THAT FLIES AROUND THE OTHER WORLD AND COMES BACK HOME WITH THE PEOPLE IN IT AND FALLS IN THE WATER

CHAIRS

COLD AIR FOR BURNING (AND BREATHING)
This part had a *very* big problem once.

PART THAT GOES ALONG TO GIVE PEOPLE AIR, WATER, COMPUTERS, AND STUFF
It comes back home with them but burns up without landing.

PART THAT STAYS ON THE OTHER WORLD
It's still there.

PART THAT FLIES DOWN TO THE OTHER WORLD WITH TWO PEOPLE INSIDE

FEET THAT GO ON THE GROUND OF THE OTHER WORLD

RING HOLDING MOST OF THE COMPUTERS

THINGS HOLDING THAT KIND OF AIR THAT MAKES YOUR VOICE FUNNY
It's for filling up the space left when they take the cold air out to burn it.

THE KIND OF AIR THAT ONCE BURNED A BIG SKY BAG AND PEOPLE DIED
And someone said "Oh, the [humans]!" (Used for burning)

PART THAT FALLS OFF THIRD
This part flew away from our world into space and hit the world we were going toward.

WET AND VERY COLD

THE PART OF AIR YOU NEED TO BREATHE, BUT NOT THE OTHER STUFF
(Used for burning)

FIRE COMES OUT HERE

PART THAT FALLS OFF SECOND

MORE SKY BAG AIR (FOR BURNING) (COLD & WET)

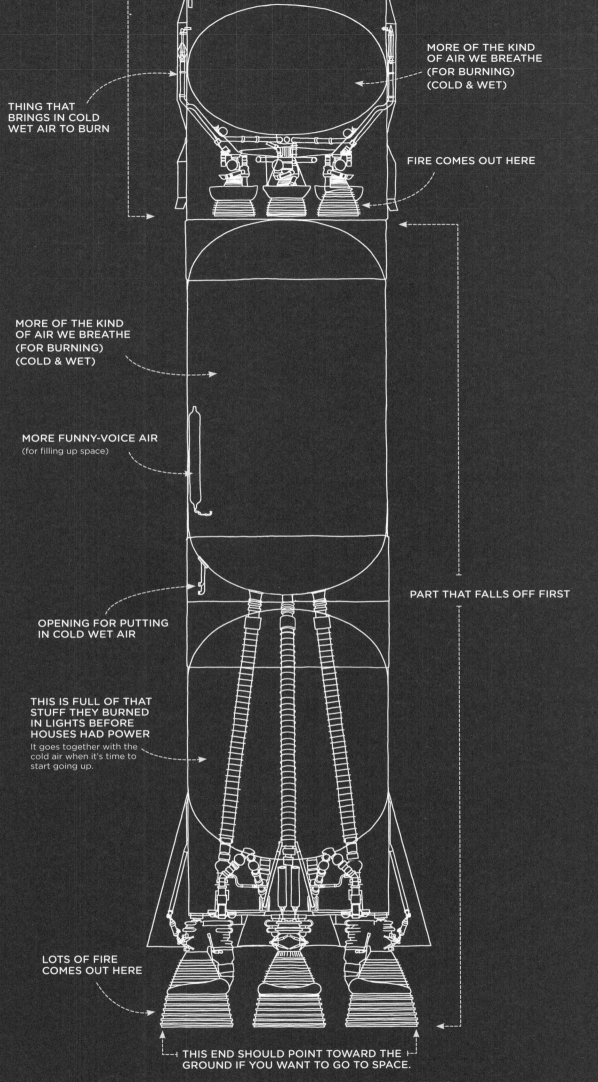

THING THAT
BRINGS IN COLD
WET AIR TO BURN

MORE OF THE KIND
OF AIR WE BREATHE
(FOR BURNING)
(COLD & WET)

FIRE COMES OUT HERE

MORE OF THE KIND
OF AIR WE BREATHE
(FOR BURNING)
(COLD & WET)

MORE FUNNY-VOICE AIR
(for filling up space)

PART THAT FALLS OFF FIRST

OPENING FOR PUTTING
IN COLD WET AIR

THIS IS FULL OF THAT
STUFF THEY BURNED
IN LIGHTS BEFORE
HOUSES HAD POWER
It goes together with the
cold air when it's time to
start going up.

LOTS OF FIRE
COMES OUT HERE

THIS END SHOULD POINT TOWARD THE
GROUND IF YOU WANT TO GO TO SPACE.

IF IT STARTS POINTING TOWARD SPACE YOU ARE HAVING
A BAD PROBLEM, AND YOU WILL NOT GO TO SPACE TODAY.

SKY BOAT PUSHER

Sky boats, like cars and sea boats, are pushed by machines that burn fire water. Fire water needs air to burn, and sky boat pushers use special blowers that use the air they're moving through to feed their fire.

Most machines that burn fire water use these four steps: **First,** pull air in. **Second,** push the air together. **Third,** burn fire water in the air, heating it and making it get bigger. **Last,** use that growing air to push on something.

Sky boat pushers use the force from the hot air in two ways: They let it fly out the back, pushing them like a space boat, but they also use it to turn their own blowers, pulling in more air and keeping themselves running.

KINDS OF PUSHERS

Small sky boats and large ones all work by pushing air, but different kinds of sky boats use different kinds of pushers.

SIMPLE PUSHER
These are fun to play with, but if you try to push any kind of boat with them, your arms get tired.

POWERED PUSHER
These are even more fun to play with (though you probably want to put them on a sky boat first).

FIRE PUSHER
These are used to push fast boats, like the kind that fight in wars. They go fast, but use more fire water than other kinds.

FIRE-POWERED BLOWER
These are like the fire pushers, but with a big blower added to the front. This kind of pusher is very good if you don't want to go too fast. They're very loud.

BIG SKY BOAT PUSHER
These are like fire-powered blowers, but they have a wall around the whole thing to control how the air goes through. They only work well when you're going slower than sound, which is why almost no big sky boats go faster than that.

HOW DO THEY WORK?

To understand how air pushers work, it can help to start by looking at space pushers.

To make a fire, you need air and something to burn. Space boats pour fire water and air into a little room that's open on one side. Then the water and air is set on fire. The fire blows up and flies out the hole, pushing the boat.

Since there's no air in space, but fires need air, a space boat has to carry air with it. Sky boats can use the air around them, so they only need to carry the fire water. They can take in air, add fire water to it, and burn it.

You can make the pusher better by using a blower in the front to force more air together into the burning room. If there's more air, the fire can burn faster and hotter.

Running the blower in front takes power. You could get that power by burning fire water in a different machine and running power to the blower with power lines. But it's better to just use a little of the power from the fire you're already making.

If you put a blower in back, in the path of the fire, it can turn a stick that turns the blower in the front. This blower slows down the burning air so it doesn't push you as well. But the blower makes the fire work so much better that it more than makes up for it.

There's one last idea that makes this work better. Instead of just using the hot air to power the blowers that press the air into the burning room, you can also use it to power a big blower.

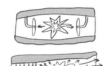

This big blower (which sometimes has a wall around it) is what really pushes the sky boat. Once you add this blower, all the rest of the parts are just there to get lots of air together, start a fire, and get power from it.

WAIT A SECOND!

One thing a lot of people wonder is "How does the force from the fire know to go out the back? Why doesn't it push on the blowers in front just as much, and slow them back down?"

The answer is that the shape of the room and the size of the blowers make it so the easier way out is through the back. It just has to push through a few blowers on the way.

STEP ONE: GET AIR
Air comes in from this side, the first step in making power.

STEP TWO: PUSH
These blowers push the air into a smaller and smaller space, which will help the fire burn faster and hotter.

STEP THREE: BURN
The air from the pushers comes into this burning room, where little drops of fire water are thrown into it and set on fire.

The fire water and air get hot and blow up. The walls make it hard to blow up in any direction except out the back, so that's where the burning air goes.

STEP FOUR: MAKE POWER
The force of the air coming out would help push the sky boat on its own, but sky boat pushers do something cooler: They put extra blowers in the path of the air. Instead of turning those blowers to push air, they let the *air* turn the *blowers*. The blowers turn the stick in the middle of the pusher, which turns all the blowers at the start, powering the machine.

That might seem like it shouldn't work, since it's using a blower to power another blower. But the power is coming from the burning fire water pushing its way out. These blowers are just a cool way to use some of that fire to keep the machine running.

BLOCKER
If there's stuff in the air, like sticks or rocks, it gets pushed through here so it doesn't hurt the blowers.

POINT
This thing helps to start pushing the air together before it goes inside.

BIG BLOWER
The fire in the back turns this big blower using the stick in the middle. This blower is what really does most of the work of pushing a big sky boat; everything else is just there to turn it.

Not all sky boats have a big blower like this. Some of them just use the hot air itself, which works well for very fast boats. But for boats going slower than sound, it turns out that using the hot air to power big blowers takes less fire water than using the air itself as a pusher.

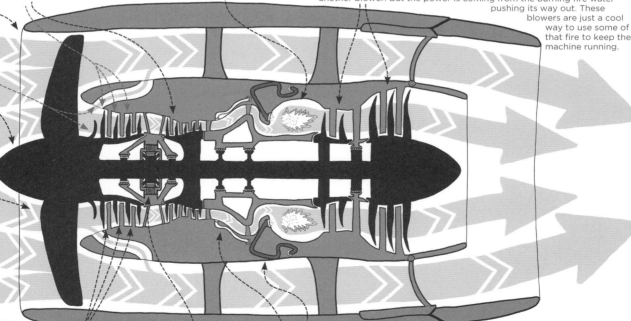

SPIN STOPPERS
The blowers that push the air together all work by spinning, but since they're all spinning in the same direction, they can start the air spinning around instead of going toward the burning room. To keep that from happening, there are little wings in between each blower to make the river of air go straight and keep it from turning too much.

POWER MAKER
This machine uses the turning stick to make a little extra power for the rest of the sky boat to use (for things like lights and computers).

FIRE-WATER CARRIERS
These carry fire water into the burning room.

AIR GETTER
The air up high is too thin to breathe. This thing grabs some of the air that the blowers pushed together and sends it to the inside of the sky boat so people can breathe.

BACK PUSHER
If the sky boat needs to stop, it can use these doors to send the air out the sides and toward the front, which makes it push back instead of forward.

STUFF YOU TOUCH TO FLY A SKY BOAT

A sky boat has a room in front with two chairs. Drivers sit in them and tell the boat where to go.

The room has some windows, but most of it is taken up by screens, keys, and little lights that change color to tell you how each part of the boat is feeling.

Sky boats have computers in them, and the computer can fly pretty well, if it has a plan for where to go. A lot of the keys and screens are for asking the boat what its plans are, and giving it new ones if you don't like what it says.

Most of the lights and keys are simple; they just do one thing, like turn a light on and off. Some of the big rows of keys and screens between the seats are for sending messages to people or looking at maps and making plans.

The hardest parts of flying are taking off and landing. The middle part of the trip is easier. During the middle part, the computer often does all the flying, and the drivers are just there to watch over it and make sure there are no problems.

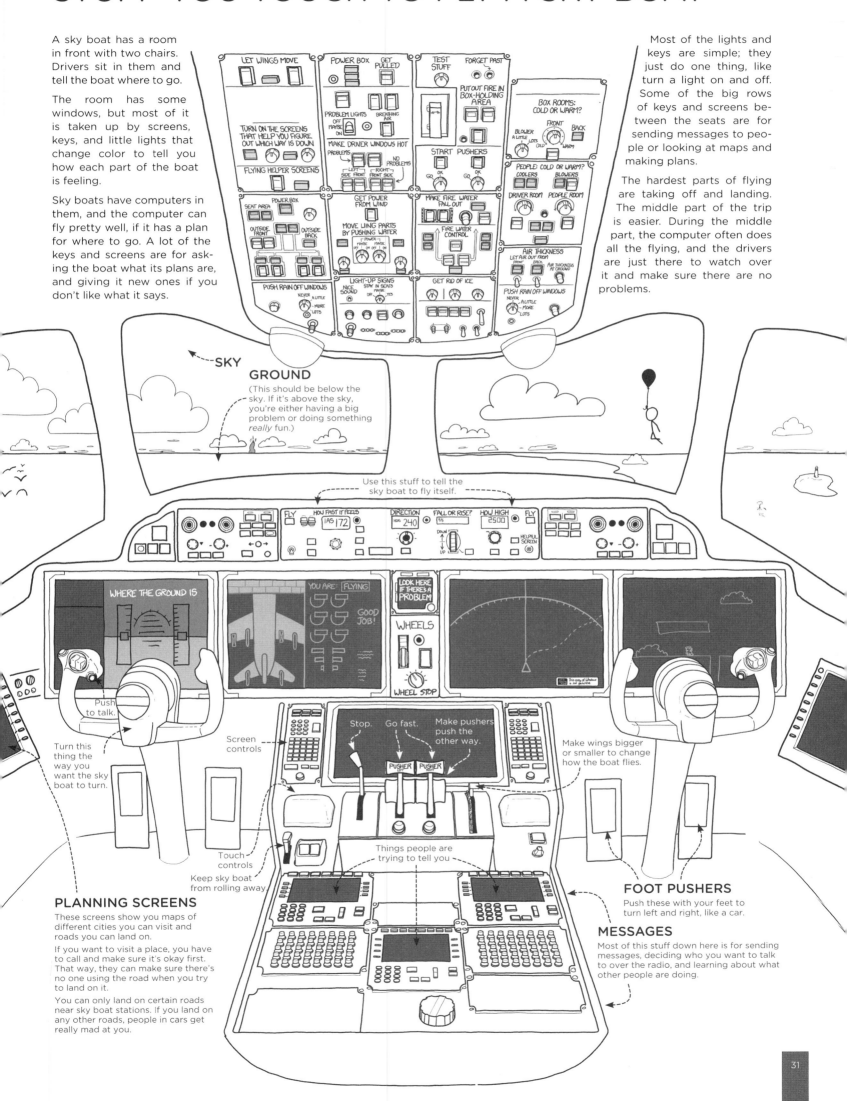

—SKY

GROUND
(This should be below the sky. If it's above the sky, you're either having a big problem or doing something *really* fun.)

Use this stuff to tell the sky boat to fly itself.

Push to talk.

Turn this thing the way you want the sky boat to turn.

Screen controls

Stop. Go fast. Make pushers the other way.

Make wings bigger or smaller to change how the boat flies.

Touch controls

Keep sky boat from rolling away.

Things people are trying to tell you

PLANNING SCREENS

These screens show you maps of different cities you can visit and roads you can land on.

If you want to visit a place, you have to call and make sure it's okay first. That way, they can make sure there's no one using the road when you try to land on it.

You can only land on certain roads near sky boat stations. If you land on any other roads, people in cars get really mad at you.

FOOT PUSHERS

Push these with your feet to turn left and right, like a car.

MESSAGES

Most of this stuff down here is for sending messages, deciding who you want to talk to over the radio, and learning about what other people are doing.

BIG TINY THING HITTER

Tiny thing hitters are machines that hit tiny things together really hard. To explain why anyone would want to do that, it might help to imagine a story about boats.

Say you and some friends are on a boat moving over a sea. The sea is covered in clouds, so you don't know what it's like. You think there's water—but what's *in* the water? Is there ice? Big biting fish? Or is it really a sea of beer instead of water? Or sand, or plastic balls?

To figure out what's in the sea, you could throw things over the edge and see what

flies back up. If you throw something heavy, a few drops of water might fly up. If you throw something down harder, the wave might toss a piece of ice into the air. You can learn a lot this way!

Now, imagine that you notice that your boat is moving. You don't have any wind sheets, so you wonder what's pushing you along.

You and your friends also notice that you sometimes hear strange sounds against the side of the boat. After thinking it over, you and your friends decide that maybe the boat is being pushed by big biting fish

hitting the side. So you get an idea: If you throw something heavy enough into the sea, one of them will throw up a big ball of water with a biting fish in it.

But to have a chance of lifting a fish into view, you need to build something that can hit the water *really* hard. Doing that will take a lot of work (and money), but you and your friends think that if it can get answers about what's going on in the water, maybe you should give it a shot.

OK, CUT THE ROPE!

SPLOOSH

BIG TINY THING HITTER
The Big Tiny Thing Hitter is the biggest and most high-power tiny thing hitter ever built. It's the size of a city, and most of it is hidden under the ground.

HOW DO WE LEARN FROM IT?
This machine works by throwing pieces of air down a hallway so they hit together really hard. The air hits with so much power that the pieces break in strange new ways, as if it shakes the air—and space itself—so hard that things fall out.

Most of these pieces only last for a moment, while space is being shaken really hard, and disappear as quickly as they appear. But by watching what flies out from the place where the air hit, we can figure out what we shook out.

WHY DID WE BUILD IT?
We're like boats, trying to understand the space we're moving around on. We can't see that space, but if we hit it hard enough, pieces fly out that tell us something.

These machines have helped us learn about space, time, and what everything is made of. We built this hitter to try to figure things out about our new ideas—about what those pieces themselves are made of, how they push on each other, and why things have weight.

WHY IS IT UNDER THE GROUND?
Space is everywhere, so we can do the hitting anywhere we want. Putting the hitter under the ground keeps it safe from things—like tiny flashes of light from space—that could make it harder to see what's going on.

START
The air starts here, in a bottle, and is pushed down this hallway.to get going fast.

HOW DO YOU PUSH AIR?
These hitters use a kind of air that can be pushed by a force from power lines, or by the kind of pulling metal used to stick pictures to a kitchen cold box. Turners are built that push the air with that force.

FAST CIRCLES
From the first hallway, the air gets sent into these hallways that go in a circle. While it's going around, they push it faster and faster.

NOT THAT DEEP
This is shown deeper than it really is, to make it easier to see. It's really only about as deep as a tall building, but as big around as a large city.

DOORS

LIFTING ROOMS

HEADING DOWN
After it goes through the top circles, the air heads down under the ground to the big circles.

FAST CIRCLE
The pieces of air fly around this hallway almost as fast as light.

WHY IT'S SO BIG
The hallway is very big; it would take you all day to walk around it. It has to be big because the air is going so fast that if they made it any smaller, they wouldn't be able to make it turn fast enough to stay in the hallway. Then the air would hit the wall and make things blow up.

PROBLEM ROOM
The flying air carries a lot of power. If they have to turn off the machine, and there's no time to let the air slow down, they send it into this huge rock room. The air hits the rock and heats it up, but doesn't hurt anything else.

HALLWAYS
The air flies through these hallways, one in each direction. To keep the fast air from hitting other air and slowing down, the hallways have all the air taken out before they start the machine. The hallways of this machine are emptier than any other place near any of the worlds around the Sun.

HITTING ROOMS
There are rooms around the hallway where people can make the air going one way hit the air going the other way. When it hits, machines in these rooms watch what flies out.

AIR POCKETS AND CLOUDS
In this machine, people watch for flying pieces using sheets of computer-controlled feelers. In older machines, though, they used some stranger things, like air pocket pools and cloud boxes.

Air pocket pools are big pools of really hot water, right on the edge of turning to air. When a tiny piece flies through the pool, it makes tiny pockets of water turn to air and start to grow. Each thing that goes through leaves a track of air pockets behind it, and they make beautiful pictures.

Cloud boxes are like air pocket pools, except instead of water right on the edge of turning to air, they use air right on the edge of turning to water. When things fly through, they leave a line of water drops in them.

You can build one of these cloud boxes in your house, and see lines left by tiny things from space! (Or from heavy metal, if you have any, which you probably shouldn't.)

AN AIR POCKET POOL

These lines are the paths of tiny things flying through water. Some of them go in circles because there's a pushing field that makes the flying pieces turn. Seeing how much they turn helps tell us what they are.

TURNERS
This machine makes the force that keeps the air in the middle of the hallway and pushes it around the circle. It works by running power through very, very cold metal. Cold metal lets power run through it very fast, and that power makes a force that pushes hard on the air.

INSIDE A HALLWAY

AIR GOING ONE WAY

AIR GOING THE OTHER WAY

COLD METAL
The metal in here is only a little warmer than the coldest that anything can be.

COOLING AIR
The outside of this carrier has a layer of air so cold it turns to water. (They use the kind of air you can breathe in to make your voice sound funny.)

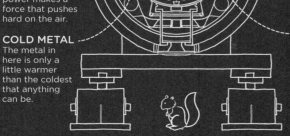

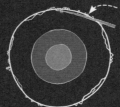

LIGHT THROUGH THE EARTH
By hitting pieces together, this hitter makes many strange things. One of the things it makes is kind of like light, except it goes right through almost anything without touching it. There's a building in another part of the world where they have machines to look at this light and learn about it.

To send the light to the building, they just point it there—right through the Earth. It's so good at going through stuff, it barely notices the Earth.

POWER BOXES

It's hard to understand how power boxes work, because they're full of water and metal doing things that are too small to see. Ideas from our normal lives don't really help us to think about what they're doing.

To try to explain how they work, we have to make up new ideas. These ideas aren't real—we can't see the "real" things—but the ideas can explain an important idea about how they work.

A lot of learning we do works this way. The ideas on this page are pretty far away from "real," but they should help explain part of how power boxes work.

IDEAS FOR THINKING ABOUT POWER BOXES

A power box has two sides, one holding a carrier wanter and the other holding a carrier maker. Between them is a wall that lets carriers through. The carrier maker would make carriers that cover the carrier wanter. But eating carriers puts pieces of power in the carrier wanter, and you can't have too many pieces of power together, because they push each other away. This stops the carrier wanter from eating too many carriers.

- ● **PIECE OF POWER**
- ○ **POWER CARRIER**
- ◉ **POWER** (in power carrier)

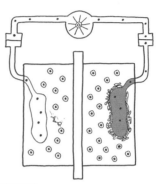

CARRIER MAKER
This metal wants to get rid of carriers. If it gets a piece of power inside it, it will send it away in a carrier made from its surface.

CARRIER WANTER
This metal wants to be covered in carriers. It will grab them if they come near and stick them to its surface, and the power from the carrier will go inside them.

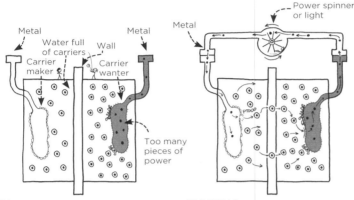

FULL
The two sides of a power box have a wall between them. This wall lets carriers through, but not pieces of power. It also stops the carrier wanter and carrier maker from touching, which would make the carriers all move to the carrier wanter without sending any power anywhere.

Extra pieces of power gather in the carrier wanter, but they can't go anywhere at first.

RUNNING
When you join the two sides with a stick of metal, the pieces of power can get from the carrier wanter to the carrier maker.

If you put a machine in their way—like a light or a power spinner—they can push on it and make it run, just like water pushing on a water wheel.

When the pieces of power get to the carrier maker, it uses them to make new carriers.

EMPTY
After a while, the carrier wanter gets covered in empty carriers, and the carrier maker gets used up. There's nothing left to push the power pieces through the metal path; the power box is dead.

With some power boxes, you can turn the wheel and push power back into the power box. This fills the power box back up.

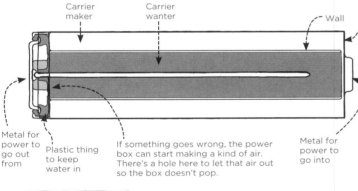

SMALL POWER BOX
This kind of power box is used in a lot of places. It powers hand lights, face hair cutters, and things kids play with.

In this kind of power box, the carrier wanter and carrier maker are made from different kinds of metal. The stuff in between is water with a kind of white stuff in it which lets the carriers move across. If the power box breaks, that stuff can come out. Don't worry, it's safe to clean up; it won't hurt your skin.

All power boxes run out of power after a while. With some kinds, you can put power back in and use them again and again, but you can't really do that with the kind shown here.

YOU CAN DROP THESE TO SEE IF THEY'RE DEAD
The carrier wanter in these boxes is made from metal dust. When it gets covered in carriers, it becomes stronger and sticks together, so the dust can't move around. This makes dead power boxes fly back up when you drop them, but full ones just hit the ground and stop.

If you cut one of these open, it would look like this—but *never* cut these open. They can blow up.

HAND COMPUTER POWER BOX
These power boxes hold more power for their size than any other. We first made them to power helper machines in people's chests. Those machines need to hold a lot of power, since people don't like it if you take them out too often.

When we started making lots of hand computers, we got better at making these power boxes, since lots of people wanted their computers to work all day without having to get power from the wall.

Of course, people also wanted their hearts to work, but more people have hand computers than heart boxes.

LIGHT METAL
In these power boxes, the carrier wanter and carrier maker are both made of very light metals. To make this kind of carrier maker and wanter work together, they're laid down in sheets almost touching each other, like two long sheets of paper laid flat and then rolled up.

CAR POWER BOX
These power boxes are used in cars. They use two kinds of heavy metal as the carrier wanter and the carrier maker, which is why they're so heavy.

Power comes in

Power goes out

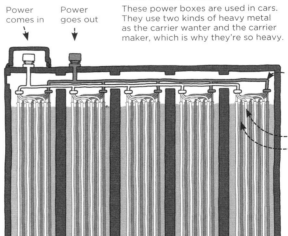

The power carrier water in between the two sides can burn your skin.

The carrier maker and the carrier wanter are two different kinds of metal, but there's something strange about this power box: When the carrier wanter gets covered in carriers, and the carrier maker makes them, they both turn into the *same* kind of metal.

HOLE-MAKING CITY BOAT

Deep in the Earth, there are pools full of the fire water and fire air that power cars and sky boats. Some of these pools are under land, and we've worked hard to get the fire water from them.

Many of the pools are under the sea floor. It's harder to reach these pools, but since you can sell the stuff in them for so much money, people have built big city boats all around the world to try anyway.

It's easy to get hurt working on a city boat. Big machines are moving heavy pieces of metal around all the time, and people work high above the water. And, of course, the whole point of the city being there is to gather stuff that burns well, so sometimes cities catch fire.

Workers on a city boat spend about half their time on the boat and half the time back on land. They usually go out to the boat for a few weeks at a time. While they're on the boat, they spend half their hours working.

CITY BOATS

Sometimes these cities sit on top of the water like normal boats, but some of them stand on the bottom on legs the size of tall buildings.

The ones in deep seas usually sit on top of the water. When they're done making holes in one area, they can drive to another.

The city uses pushers to stay over the hole. They're like the pushers on a normal boat, but bigger.

If there's a problem and the city blows away or catches fire or falls over or something, this stick joining it to the hole can break off.

If it does, this part keeps the fire water from coming out.

People don't hear very much about this part when it works. When it doesn't work, they hear about it a *lot*.

SEA FLOOR

FIRE WATER POOLS

HOLE

ROCK-EATING STICK MACHINE

We push this stick into the ground, and wheels at the front covered in teeth turn to cut the rock into tiny pieces and push it back up the hole.

METAL STICK

HOLE

Water being pushed down to the end of the stick to help carry broken rocks away

BENDING PART

We used to make our holes straight down, because anything else was too hard to keep track of.

Now, with the help of computers, we can point rock-eaters very carefully even deep under the ground. This lets us make longer holes with many branches and get more stuff out of the ground.

Water coming back up carrying broken rocks

Teeth for breaking rock

Machine for feeling how much the stick is turning so the computer can tell it where to go. (There's one of these in your phone to tell which way you're holding it.)

ROCK

FIRE WATER GETTERS

These things get the fire water either by pulling it up with a metal stick or pushing it up with a machine.

Sometimes the fire water comes up on its own because of the weight of the Earth pushing down on the hole it's in. Most of the time, that's nice, but sometimes it can be a very, *very* big problem.

PULLING STICK

Fire water getting pulled up

Fire water getting pushed up

POWER LINE

PULLER

FIRE WATER POOL

SPINNING WATER PUSHERS

These are like the things you use to lift water out of the bottom floor of your house (or move water around a glass fish holder). They spin around and blow the fire water up to the surface.

ENGINE

FIRE WATER POOL

FIRE WATER POOL

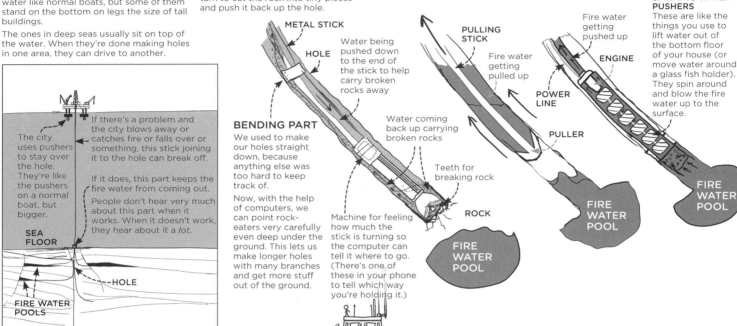

PLAYING FIELD

Some city boats have fields for playing games. They have to put walls around them, though, because if you kick your ball over the side, you can't get it back.

GAME ROOM

There are usually rooms with different games in them, for workers to play while they rest, like a kind of circle-stick ball played on a table.

Room where fire water comes up

Machine for lifting fire water, sand, and rock-eating machines out of the hole

These things sit on top of the hole and control what kinds of air and water go in and out.

They're named after those winter family trees (the kind with boxes under them) because people say they look like one.

LIFTER

LIFTER

VISITING SKY BOAT

CONTROL ROOM

FISH WATCHER

Some workers like to stand near the edge and look for cool fish.

DOCTOR ROOM

Room for fixing boat parts by sticking things together with fire and power and cutting things with glass-stone teeth

Room for drinking warm things and waking up

ENGINE ROOM

FOOD ROOM

SLEEPING ROOMS

CLOTHES WASHERS

STORE

MOVIE ROOM

Rock-eating metal sticks go down through this thing, and sand, rock, and fire water come up.

WEIGHT ROOM
Room for lifting things and running in place to get stronger

VISITING BOAT

FLYING FISH

LIGHT THINGS TO LET THE CITY STAND ON WATER

WATER

TO SEA FLOOR

BITING FISH

PUSHERS

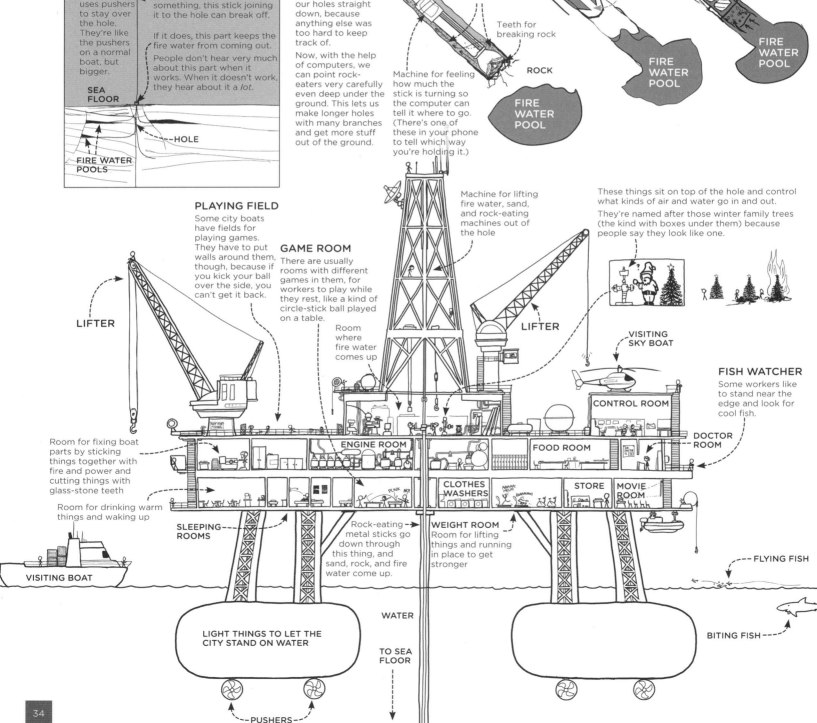

STUFF IN THE EARTH WE CAN BURN

Almost all living things are powered by the Sun. Some living things get their power straight from the Sun's light—like trees, and some things that grow in the sea. Most living things that don't eat the Sun's light eat other living things to get *their* power. In the end, the power comes from the Sun.

When things die, some of that power is left in their remains, which is why you can get power out of dead trees by burning them.

Sometimes, if dead things don't burn or get eaten, they go into the ground with that power still inside them. Over a long time, under the weight and heat of the Earth, huge numbers of these remains can change into different kinds of rocks, water, or air . . . but even as they change, they hold on to their power. When we find these remains, we can burn them, and get all that power—gathered from the Sun over huge stretches of time—at once.

When we first built machines powered by fire, we burned wood from the forests of our time. When those weren't enough, we started burning the forests of the past.

One day, those will run out, too, and we'll have to get power somewhere new—like straight from the Sun, or the Earth's heat.

But we may have to change the kind of power we use soon, before we finish burning all the stuff in the ground. It turns out burning that stuff is changing our air, in a way that's making the world hotter. If we use up all the black rocks, fire water, and fire air, the problem it makes may be too big for us.

HOW WE GET BLACK ROCKS OUT OF THE GROUND

THEY DO THIS IN THE MOUNTAINS NEAR WHERE I PLAYED AS A KID.

If the rocks aren't very deep, we can make holes under the ground and carry them up with machines. This is how we used to get most of the rocks we burned.

As we built bigger earth-moving machines, we learned to just move all the trees and land out of the way to get the rocks.

Some rocks are inside mountains, so some companies have started blowing up the tops of the mountains so they can get the rocks out more easily.

This kind of work leaves pools full of heavy metals and strange kinds of water that was used to get the black rocks out. Sometimes you can notice the bright colors of these pools from the air. When companies are done making holes, they often leave the pools behind. People worry about whether the stuff in the pools could be bad for us. Sometimes birds land in the pools and die.

HOW WE GET FIRE WATER AND FIRE AIR OUT OF THE GROUND

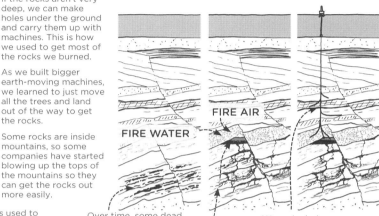

FIRE AIR

FIRE WATER

Over time, some dead things slowly turn to fire water and fire air.

These are both lighter than rocks, and rise up through tiny holes. When they reach a rock with no holes, they form pools, with the lighter air on top.

We make holes looking for places where lots of things died. When we find a pool, we push a stick down and pull up all the fire air and fire water.

BLACK ROCKS

HOLES

One reason we make holes that bend is so we can reach under cities without bothering people.

HOW DEEP?
We can only get black rocks easily if they're not too deep in the ground. The biggest problem is that deeper in the Earth, rocks are hotter. It's hard to get a lot of rock up out of the ground, and if the rocks are too hot, that makes everything so hard that it's not worth it.

There are other problems. You need to cut big rooms into the ground to get black rocks out, and it's hard to hold the roof up when there's so much rock piled up on it. Sometimes the roof falls and people die.

STRANGE SHAPE
When a sea dries up, it leaves lots of this white stuff behind. Sometimes, the stuff gets covered in dirt and sand.

When the layers above the white stuff get heavier, it can make the white stuff start to rise up and push through the layers above. It looks like paint drops falling from a ceiling, but going up.

WHITE STUFF
This is white stuff, like what we put on food to make it better (although we mostly get the kind we eat from drying out sea water). We make holes like this to get white stuff out, then we put it on our roads to get rid of snow and ice.

We sometimes use the spaces we leave behind to hold stuff, like fire water or fire air that we want to save to burn later.

Layers of rock from different times

HOLES

FIRE WATER

Places where the ground broke

DEEP POOLS
We can get fire water and fire air from much deeper places than we can get black rocks. Since it forms pools and can run through small holes easily, we only need to make a very thin hole to get it out, instead of having to move all the rocks around it.

FIRE WATER

FIRE AIR
(on top of the fire water)

ROCK BREAKING
Big, easy-to-reach pools of fire water are getting harder to find, so we've been trying new ideas for getting it from the ground. We've found that sometimes, rock has fire water or air you can burn stuck in it. To get it out, we push water into the ground so hard that it makes the rocks break. Then we push in small rocks or glass to hold the breaks open, and the fire water and fire air comes out through the openings.

Making all these holes in the rock might mean that when we drink water, we'll also drink whatever stuff they use to get fire water out, since everything can run through the new holes in the rock.

VERY DEEP HOLES

TALL ROADS

The Earth's pull holds people to the ground. We like to walk around, but sometimes the ground goes places we don't want to go, like under a river or into a deep hole. We can't get past those places because we have to follow the ground. (Birds don't, since they can fly by pushing on the air. Someone in a movie once sang, "If birds fly over the sky, why can't I?" The answer is, "You are too big and don't have wings.")

If we want to go somewhere, we can make a road that goes straight across, high above the ground. Making short roads over holes and rivers is pretty easy, but making long ones can be very hard.

HOLE

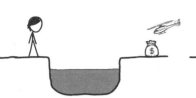

Sometimes, you want to walk somewhere, but you don't want to go where the ground goes.

ROAD

If the hole is small enough, you can put a board over the hole to make a new road. Then you can walk across the board.

LONG ROAD

If you find a bigger hole, you can try to find a bigger board. Bigger boards are longer and stronger, but they're also heavier—and bigger things get heavier faster than they get stronger.

LONGER ROAD

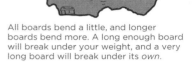

All boards bend a little, and longer boards bend more. A long enough board will break under your weight, and a very long board will break under its *own*.

BENDY ROAD

You can cross a larger hole with a road that's allowed to bend. If you tie many small boards together and let it hang, bending won't hurt it, and that will let it hold more weight.

This kind of road gets stronger the more you let it hang down, but it also gets harder to walk across. If it hangs down *too* far, it's no better than just walking down into the hole.

THICK ROAD

You can cross a bigger hole if you make the board thicker. Thicker things are harder to bend, so this kind of road is stronger.

TALL ROAD

It might seem like it would make more sense to put the extra-thick part below the road, because it's "holding" the road up, and we usually hold things from below.

But if it's strong mostly because it's thick, then it works just as well if you add the thickness above the road.

ROAD HANGING UNDER A STRONGER SHAPE

Since all this stuff you're adding is just there to hold the road up, that stuff doesn't need to be near the road. You can make a strong metal piece that goes high up over the hole—which gives it a stronger shape, but would be harder to walk on if the road went that way—and then use strong metal lines to hang the road straighter across under it.

ROAD HANGING FROM STICKS

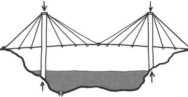

Another way to hold up a road is to build very strong sticks, then hang the road from the top ends of the sticks. The lines need to be a little stronger than the lines in the other hanging road, and the sticks need to be *really* strong. On the other hand, there are only two sticks, so that can make building easier.

HANGING ROAD PROBLEMS

When you hold up a road by hanging it, you have to be very careful. These tall roads keep the road from being moved by the Earth's pull—which is always straight down—but wind can make the road swing side-to-side.

Some roads have fallen down because the builders didn't understand wind well enough.

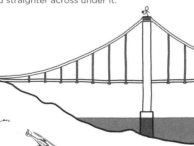

Strong line
Smaller lines

VERY TALL ROAD

THIS IS THE BEST KIND OF TALL ROAD.

That's not really true. Different tall roads are good for different things.

But a lot of the time, when you need to cross a big hole, this kind of shape will let your road reach farther than any other shape would.

TALL ROADS ON OTHER WORLDS

A very bright person (who was known for calling the Earth a "pale blue point") once said something interesting about these kinds of roads in one of his books.

He pointed out that everything about the shape of very tall roads is decided by the laws of space and time—the laws that say how a world's weight pulls things—and those laws are the same everywhere.

That means that if there's life on other worlds, the road shape that works best for them should be the same one that works best for us. Our tall roads may look familiar to them.

Maybe that's true; maybe it's not. We don't know if there's life on other worlds, and if there is, maybe they don't build roads at all. Maybe their way of living is different from ours in ways we can't even think about.

But if they have holes they need to get across . . .

. . . and if, in their world, they build things out of different shapes, like us . . .

. . . and if they have problems with holding their roads up . . .

. . . then they very well may build tall roads that look just like ours.

I like that idea, because now, when I look at one of these tall roads, I always feel a little happier. It makes me think about how maybe, somewhere far across space and time, there's someone looking at another tall road, thinking about how the shape might be found across many worlds, and—maybe—wondering about me.

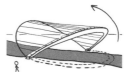

BENDING COMPUTER

HOW TO OPEN A COMPUTER

HOW TO OPEN A COMPUTER EVEN MORE
Different computers open differently. Some open from the top, by lifting the keys away, and some are really hard to figure out.

Note: If you're trying to use your computer, you probably don't need to do this. Just open it the normal way, like in the first picture.

If you open it like this, you might not be able to use it for a while, or maybe ever.

BENDER

KEYS
You can press these to put words in the computer.

SCREEN

OUTSIDE
(plastic or metal)

TOUCHING BOX
You can touch this to point at things on the screen. Some computers can feel when you touch the screen itself, so you don't need to use this box—but even on those computers, using the box can be be easier because you don't have to lift your hand.

BOTTOM PIECE

TURNING HOLDERS
(Don't lose these.)

STICKERS THAT YELL AT YOU
These are put here by the company that made the computer. They tell you that if you open the computer and then it breaks, they don't have to fix it.

INSIDE

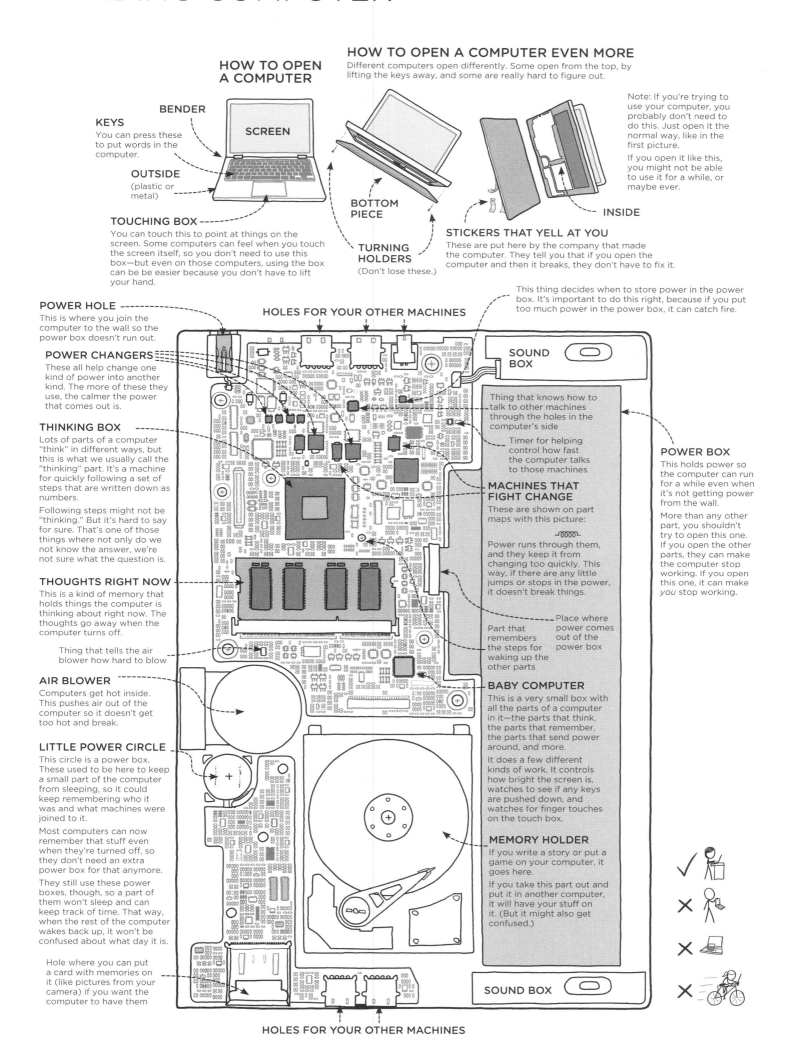

POWER HOLE
This is where you join the computer to the wall so the power box doesn't run out.

POWER CHANGERS
These all help change one kind of power into another kind. The more of these they use, the calmer the power that comes out is.

THINKING BOX
Lots of parts of a computer "think" in different ways, but this is what we usually call the "thinking" part. It's a machine for quickly following a set of steps that are written down as numbers.

Following steps might not be "thinking." But it's hard to say for sure. That's one of those things where not only do we not know the answer, we're not sure what the question is.

THOUGHTS RIGHT NOW
This is a kind of memory that holds things the computer is thinking about right now. The thoughts go away when the computer turns off.

Thing that tells the air blower how hard to blow

AIR BLOWER
Computers get hot inside. This pushes air out of the computer so it doesn't get too hot and break.

LITTLE POWER CIRCLE
This circle is a power box. These used to be here to keep a small part of the computer from sleeping, so it could keep remembering who it was and what machines were joined to it.

Most computers can now remember that stuff even when they're turned off, so they don't need an extra power box for that anymore.

They still use these power boxes, though, so a part of them won't sleep and can keep track of time. That way, when the rest of the computer wakes back up, it won't be confused about what day it is.

Hole where you can put a card with memories on it (like pictures from your camera) if you want the computer to have them

HOLES FOR YOUR OTHER MACHINES

This thing decides when to store power in the power box. It's important to do this right, because if you put too much power in the power box, it can catch fire.

SOUND BOX

Thing that knows how to talk to other machines through the holes in the computer's side

Timer for helping control how fast the computer talks to those machines

MACHINES THAT FIGHT CHANGE
These are shown on part maps with this picture:

Power runs through them, and they keep it from changing too quickly. This way, if there are any little jumps or stops in the power, it doesn't break things.

Part that remembers the steps for waking up the other parts

Place where power comes out of the power box

POWER BOX
This holds power so the computer can run for a while even when it's not getting power from the wall.

More than any other part, you shouldn't try to open this one. If you open the other parts, they can make the computer stop working. If you open this one, it can make *you* stop working.

BABY COMPUTER
This is a very small box with all the parts of a computer in it—the parts that think, the parts that remember, the parts that send power around, and more.

It does a few different kinds of work. It controls how bright the screen is, watches to see if any keys are pushed down, and watches for finger touches on the touch box.

MEMORY HOLDER
If you write a story or put a game on your computer, it goes here.

If you take this part out and put it in another computer, it will have your stuff on it. (But it might also get confused.)

SOUND BOX

HOLES FOR YOUR OTHER MACHINES

DARK WORLDS

Out past the cold wind world, there are lots of little ice worlds moving very slowly around the far-away Sun.

BIG TRIP TAKER ONE

We were so surprised when an earlier boat saw clouds around the cloud moon, we told Big Trip Taker One to change plans and fly past the cloud moon for a closer look. This took it off the road to the other worlds, and it headed out into space. It's now traveled farther from home than anything humans have built.

BIG TRIP TAKER TWO

Big Trip Taker Two is the only boat to visit the two outer worlds.

This world is far from the Sun. It has the coldest air and the fastest winds.

BIG WORLD BOAT

This boat visited the big world and its moons. Once it was done with its job, we told it to fly into the huge air world so it would burn up in the air, just like old space boats sometimes do on Earth.

We did that because we were worried that if we didn't, it would hit one of the other worlds and spread Earth's tiny animals there. We don't know if there are any other animals on those worlds, but if there are, we don't want our animals to eat them before we can look at them.

These two air worlds are smaller than the ring world and the huge air world. They have more kinds of water in their air, which makes them more blue.

LOST WORLD

This world is strange because it used to go around the Sun alone, but one day it came too close to the cold wind world and now it lives there.

OLD CIRCLE-COVERED MOON

This moon got hit by lots of rocks once, which left it covered in circle-shaped holes, just like our Moon.

These two worlds are huge balls of air and water with some rocks in the middle.

RING WORLD

All the big air worlds have thin rings, but this world's are huge and bright.

RING WORLD BOAT

This boat visited the ring world to learn more about it—and to get a closer look at the cloud moon.

BIG MOON

This is the biggest and heaviest moon near the Sun.

BIG WORLD

This is the biggest world around the Sun. It's mostly made of air. Some of its moons are almost as big as our world.

CLOUD MOON

This world is very strange—it's the only moon covered in thick clouds. The air there is even thicker than Earth's.

It would be nice if it were the kind of air we can breathe, but it's not.

SMELLY YELLOW MOON

This world has lots of color, but the color doesn't look very nice. It looks kind of like fire, but even more like food that came out of someone's mouth. It's covered in stuff that smells like old food.

ICE WATER MOON

This world has ice on the outside, but it's warmer on the inside, and there's water under the ice.

Since there's warm water, lots of people want to go there to look for animals. We don't know if there are animals there, but if there are, we want to know about them.

SPECIAL ENGINE BOAT

This boat visited two of the little worlds between the red world and the huge world. It's pushed by a special engine that's powered by the Sun.

It was the first boat to visit two different worlds and stay for a while at each of them.

THE MOON

Other moons have names, but our moon is just called the Moon. Some people visited it once.

We're not sure where it came from. We think another world probably hit our world when it was very young and lots of pieces flew off, and then the pieces fell together and made the new baby world. But we don't know for sure yet.

LITTLE RED WORLD

RED WORLD SPACE CAR

HOT SKY WORLD

This world is about as big as our world, but it's much hotter. One reason it's hot is that it's closer to the Sun. The other reason is that it has more air than our world, which keeps it warm, sort of like a thick coat around the whole world.

People used to think this world would be nice to live on. But if you visit this world, you will have lots of problems.

The air is really hot. If you land there, you will be on fire, and you will not come home. The air is really heavy. If you land there, it will be like you're deep under the sea. The sky will press down on you and make you get smaller, and you will not come home. The air isn't the kind of air humans breathe. If you try to breathe it, you will not come home. The air is also full of a kind of water that's bad for your skin. If it touches you, you might come home, but without your skin.

OUR WORLD

There are animals and trees and a blue sky here. You probably live here, too; it's very hard to leave.

SMALL ROCK WORLD

This world is hard to see because it's right next to the bright Sun. It turns very slowly, so the day side gets very hot and the night side gets very cold.

THE SUN

The Sun is a star. It looks bigger and brighter than other stars because it's much closer to us. But it turns out it really is bigger and brighter than other stars.

For a while, we thought it was smaller than other stars, because most of the stars we looked at turned out to be bigger than it. But it turns out there are lots of stars that are less bright; they're just harder to see.

WORLDS AROUND THE SUN

The Sun is the biggest thing near us. Our world and everything near it goes around the Sun. Some of the worlds that go around the Sun are big enough that they have their own moons—little worlds that go around them as they all go around the Sun.

All of our history happened in this picture. Most of it happened on the third world, counting out from the Sun. You're somewhere in this picture right now!

. . . probably. But sometimes, books last a very long time. Maybe you're reading this hundreds of years after I wrote it. Maybe you're on a boat or a world that's somewhere outside this picture.

If you are, then I'm wrong. But I'm happy to be wrong for such a cool reason! I just wish you could tell me what you've seen.

THE LITTLE RED WORLD

This world had seas on it when it was very young, but now it's cold and the seas are gone.

It's called the red world because there's metal in the sand, and it turned red over time, for the same reason that old keys or trucks turn red on Earth if you leave them outside for a long time.

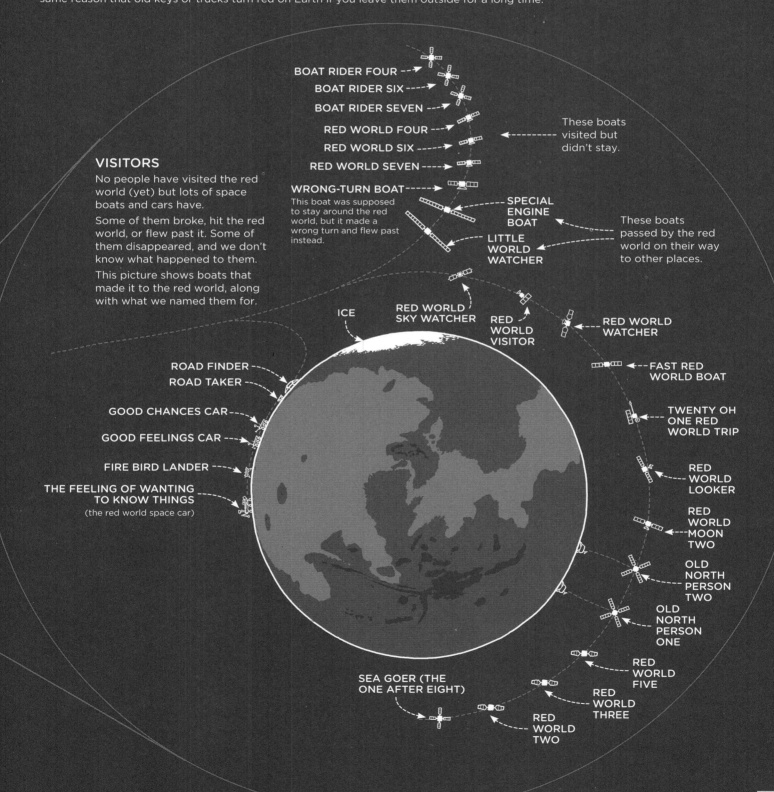

VISITORS

No people have visited the red world (yet) but lots of space boats and cars have.

Some of them broke, hit the red world, or flew past it. Some of them disappeared, and we don't know what happened to them.

This picture shows boats that made it to the red world, along with what we named them for.

BOAT RIDER FOUR
BOAT RIDER SIX
BOAT RIDER SEVEN
RED WORLD FOUR
RED WORLD SIX
RED WORLD SEVEN
WRONG-TURN BOAT

This boat was supposed to stay around the red world, but it made a wrong turn and flew past instead.

These boats visited but didn't stay.

SPECIAL ENGINE BOAT
LITTLE WORLD WATCHER

These boats passed by the red world on their way to other places.

ICE
RED WORLD SKY WATCHER
RED WORLD VISITOR
RED WORLD WATCHER
FAST RED WORLD BOAT

ROAD FINDER
ROAD TAKER
GOOD CHANCES CAR
GOOD FEELINGS CAR
FIRE BIRD LANDER
THE FEELING OF WANTING TO KNOW THINGS
(the red world space car)

TWENTY OH ONE RED WORLD TRIP
RED WORLD LOOKER
RED WORLD MOON TWO
OLD NORTH PERSON TWO
OLD NORTH PERSON ONE
RED WORLD FIVE
RED WORLD THREE

SEA GOER (THE ONE AFTER EIGHT)
RED WORLD TWO

PICTURE TAKER

When you look at something, the light from it goes into your eye and makes a picture inside your head. The picture gives you an idea about the thing's shape and color.

Since before humans learned to write, we've been using painting to turn our ideas back into pictures. Pictures let us remember the things we saw and ideas we had, and to put those ideas in other people's heads, too.

A few hundred years ago, we started making machines that turned light straight into pictures. These made it easier for anyone to make pictures, and picture making has become a big part of how we talk and share.

OKAY, SMILE!

LIGHT PAPER

Some kinds of paper change color when light hits them. Picture takers used these for a long time.

This paper alone isn't enough to make a picture, though. When you hold the paper up to someone, light from every part of them hits every part of the paper, so your whole page will be all one color. (Unless you hold the paper so close to the thing that each part of the paper only sees light from one part of the thing, but that doesn't work very well.)

I HATE HOW EVERYONE TAKES PICTURES INSTEAD OF JUST ENJOYING THE VIEW.
...YOU SAY, INSTEAD OF ENJOYING THE VIEW.

SHAPE

To make a picture of something, you need to control the light so that each part of the paper sees light from just one part of it.

One way to do this is by blocking almost all the light paths using a wall with a hole in it. (This makes a picture that's turned over, but that's okay—you can just turn it back.)

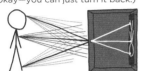

MORE LIGHT

The hole idea works, but a tiny hole doesn't let very much light through, so it takes a long time for enough light to hit the paper to make a picture.

To let in more light, you could make the hole bigger, but then the light from one spot starts to spread out on the paper, clouding the picture.

BENDING LIGHT

To make the picture less clouded, we need to bend lots of light from each part of the thing toward the spot on the picture that goes with it.

We can do this by using things that bend light—like water and glass.

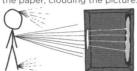

SPECIAL SHAPES

By cutting glass into the right shapes, we can make light benders that catch lots of light and send the light from each direction to a different part of the picture.

This machine is good enough to take a simple picture, but it will be a little clouded and not very sharp or bright. To take a clearer picture, we have to add more benders to control the path the light takes more carefully.

Most picture takers use glass, since it's easier to cut it into a shape than water. Some people are trying to build computer-controlled benders that use water, which would let the benders change shape to control the light without using as many parts.

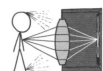

NOT AGAIN...

BIG PICTURE TAKER

This machine is used to take sharp pictures, even of things that are small or far away.

Our eyes are better than most picture takers at seeing small and far-away things, but thanks to its very large benders that take in lots of light, this kind of picture taker can see even better.

WHY ARE THERE SO MANY BENDERS?

These light benders are here for different reasons, but one of the big ones is that some colors of light bend more than others when they go through glass. This can make some colors in a picture sharp while others are spread out. Different kinds of glass break up colors in different ways, so by sending the light through one kind of glass and then another, groups of benders can get the different colors to the same place.

POWER BOX

Taking pictures can use lots of power, so picture takers usually need special power boxes.

MEMORY

This holds the pictures you take.

LOOKER

The whole front end of the picture taker is for gathering light. The whole thing can come off, so you can use different lookers for different kinds of pictures.

PICTURE WINDOW

This window opens and closes to let light through to the light catcher and take a picture. It has two sheets. When it starts taking a picture, the bottom sheet pulls down out of the way. When it's done gathering light, the top screen comes down to cover it. It uses two screens; if it used a screen that came up and then pulled back, then the top half of the light catcher would spend more time catching light than the bottom.

LIGHT CATCHER

This used to be made of paper, but on computer picture takers like this one, it's a flat sheet of computer light feelers. Each one checks how much light is hitting it, then tells the computer. The computer puts the messages together to make a picture.

SCREEN

This screen shows you what the light catcher is seeing. It also lets you look at the pictures you took and decide whether you want to keep them.

Some picture takers have a hole you can look through, too, which shows you a view out the looker using a mirror (or pretends to, using another screen).

FLASH

If there's not enough light to make a good picture, this can light up the area for a moment while the picture window is open. The light can make the shadows in a picture look strange, though, so some people try not to use it very much.

LIGHT COMES IN HERE

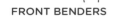

FRONT BENDERS

These grab all the light and start bringing it together so the other benders can do things with it.

CLOSE-OR-FAR BENDERS

These benders control how close up or far away the things in the picture look. They slide forward to look at small things far away, and back to see a wider view of the whole area.

DUST SHAKER WINDOW

Even a tiny piece of dust stuck to the picture window can make the machine take bad pictures. In small picture takers, the picture window is locked inside and safe from dust. But on ones where you can take the big looker off to put on a different one, dust can get in.

To keep dust from being a problem, there's a window in front of the picture window with a shaker on it. The shaker shakes the window very fast, throwing off any dust that sticks to it.

PICTURE BENDERS

These light benders are the ones that bring the light together to make a picture on the light catcher in the back.

NO MIRROR

Nice picture takers used to have a mirror here, so you could look through a hole in the top and see through the looker, to see what would be in the picture. The loud "picture-taking sound" is the mirror moving out of the way to let light reach the back.

Now, more and more picture takers are using screens to show you the view instead.

CHANGING SHAPE

Picture takers have changed shape over time. The back parts are smaller, but some of the front parts of good picture takers have stayed big. The jobs done by the back parts, like saving pictures and storing power, are now being done by small computers. The front parts bend light, and computers can't do that yet.

Soon, people might just use their hand computers as the back part, sticking them to a looker to take nice pictures.

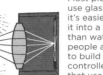

FROM YOUR PICTURES, WE FIGURED YOU WERE JUST BAD AT USING CAMERAS.

NO, I'M ACTUALLY LIKE THIS.

WRITING STICKS

People used to write all their words using sticks. Now, they write words by pressing keys, which is usually a lot faster. Even though we're writing more words every day than we used to, we're using writing sticks less and less.

Some people still use sticks like these for things other than writing. People who draw pictures for a living might not use paper anymore, but most of them still use sticks to control where lines go.

(The pictures in this book weren't drawn on paper, but they were still drawn with a stick.)

Someday, we may stop using sticks for making pictures, too.

WOOD KIND
(makes lines you can change)

PROBLEM FIXER
This used to be made from tree blood, but now it's usually made from a kind of plastic (which is made from very old dead things).

WOOD PART
Most of the wood for most writing sticks comes from one kind of tree. Writing-stick makers like that tree because the wood smells nice and doesn't break off into little pieces that get stuck in your hand.

WRITING PART
This used to be made from a kind of heavy metal, long ago, but now it's made from black rocks that we found in the ground. Most of those rocks come from very old dead things.

It turns out a lot of the parts of writing sticks come from dead things.

POINTS
This kind of writing stick gets less pointy as you use it, so you have to keep cutting it to make a new point.

Sometimes you get a nice sharp point, but sometimes it's hard to get right:

When this happens, it can make you decide the thing you were writing wasn't very important after all.

WATER KIND
(makes lines you can't change)

If you write something wrong with these, it will always be wrong. (Unless you write over it with the same color as the paper, so it's hidden.)

SOUND
You push down on this part to make the end of the writing stick come out.

It makes a loud sound when you do this, and some people press it over and over again.

They don't always think about the sound when they're doing this, but the people around them do.

CLICK
CLICK
CLICK

YELLOW PAINT
Although we haven't used heavy metal in the writing part in the center for hundreds of years, until not too long ago, we *did* use it to make the yellow paint on the outside.

We finally learned how bad that metal was and stopped using it for yellow paint, so now you probably won't get sick from biting your writing stick.

Empty space where you can hide a piece of paper if you want to send someone a note

PUSHER
This is the bottom end, where the writing water comes out slowly.

Sometimes it comes out very fast, and colors your clothes. Then you have to buy new clothes, unless you like the new colors.

HOW THE END THING YOU PRESS WORKS
If the bottom end of the writing stick, where the writing water comes out, is inside the stick, pressing the top end makes it go out. If it's out, pressing again makes it come back in.

The writing stick does this by using several strangely shaped parts that all push on each other in different ways, and it's hard to understand even when you can see it happening.

It's a very cool idea.

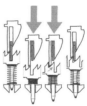

BALL MADE OF VERY, VERY HARD METAL
The ball at the bottom end rolls around as the writing stick moves over the paper.

The top side of the ball gets covered with writing water, and then it rolls over and leaves the water on the paper.

END HAT
This is made of plastic, like most of the parts of this writing stick. If you take this off, the writing water won't fall out, but it will make the writing stick look less nice.

WRITING WATER
This is the stuff that goes on the page. We used to get it from sea animals that have many arms.

Now—like almost every other part of these sticks—writing water is often made from black rocks and fire water, which are themselves made from long-ago dead trees and sea life.

Some writing water, like the kind used to write news on paper each day,* is made from a kind of food we grow (it's a kind of food people use to make pretend animal parts to eat instead of real ones).

*If you're reading this in the future, this is a thing we used to do.

ANIMAL WE GOT WRITING WATER FROM

HUMAN

	HUMAN	
ARMS	2 +2 OTHERS	8 +2 OTHERS
BONES	206	0
BIG BRAIN	✓	✓
WRITING WATER BAG		✓
CAN DRIVE CARS	✓	
STRANGE	✓	✓

When the ball dries out, you have to move it around for a while before it gets covered with writing water again.

hello

HAND COMPUTER

These machines began as radios for talking out loud to people who were far away. Over the years, they slowly became more and more like computers.

As these machines turned into computers, they started taking the place of a lot of things we used to carry around—like picture takers, music players, and even books.

FACE CHECKER
This turns the screen off if your head is near it, so you don't press keys with your face while you're talking.

FRONT CAMERA

SOUND MAKER

BIG CAMERA

EAR PIECE HOLE

POWER KEY
You can push this to make the computer sleep or wake up.

HOT SPOT TALKER
This lets the hand computer talk to people through a little radio in your house, instead of the big radio run by the phone company, which can save you money.

EXTRA MEMORY HOLDER
If your phone is storing too many memories for you (like pictures, sounds, and games), you can put a card here to give it more space.

As computers and radios get faster, companies are holding more and more of your memories on their computers, and only sending them to you when you ask for them.

LIGHTS
For taking pictures

POWER BOX JOINER

POWER BOX

POCKET MOVER
This piece of metal turns very fast to make the phone move. That way, it can get your attention without making too much noise. (Unless it's sitting on a hard table; then it can make a *lot* of noise.)

JOINERS
Different parts of the hand computer, like the screen and the radio feelers, join up with the rest of the phone here when it gets put together.

CARD HOLDER
This holds the card that lets the phone talk to the world. The phone works by using a radio to talk to a company that you pay to carry your messages. It uses this card to tell them which hand computer they're talking to.

SOUND UNDERSTANDER

RADIO TALKER
This tells the hand computer how to understand the words the company's radios send.

THINKING BOX

DIRECTION FEELER

FAST MEMORY
This part of the phone holds stuff the hand computer is thinking about right now, like pages you're looking at or games you're playing. The memory goes away when the phone turns off.

BIG SOUND MAKER
This thing makes noise that you can hear even when your ear is far away from the phone.

LOUD CONTROLS
These controls make the sound in your ear louder or quieter.

TINY POWER GATE
Like other computers, almost every part of a hand computer is full of many different kinds of power gates.

This picture is used in maps of parts to mean "power gate":

These gates take in power from one line, and listen to another line to decide whether to let the power through or not. Computer brains are built by sticking these gates together.

There are as many power gates in a computer as there are people on Earth. Some of them are big and easy to see, but most of them are tiny and control very little power. The gates, I mean, not the people.

LISTENING BOX
This is a special thinking box that just listens for words. Since it only does one thing, it can do it without as much power as the main thinking box would need. If a phone has this, you can make it listen for your voice all the time, not just when you press a key.

RADIO FEELER
This part listens to the thin pieces of metal along the outside of the hand computer. When a radio message comes in, it makes power move in the metal. This thing listens to how the power is changing and turns it into words.

It also listens to words that the hand computer wants to send back out, and turns them into power changes to send down the metal.

POWER HOLE

POWER CONTROLLER
This thing pays attention to what different parts of the computer are doing, and makes sure each part gets sent the power it needs.

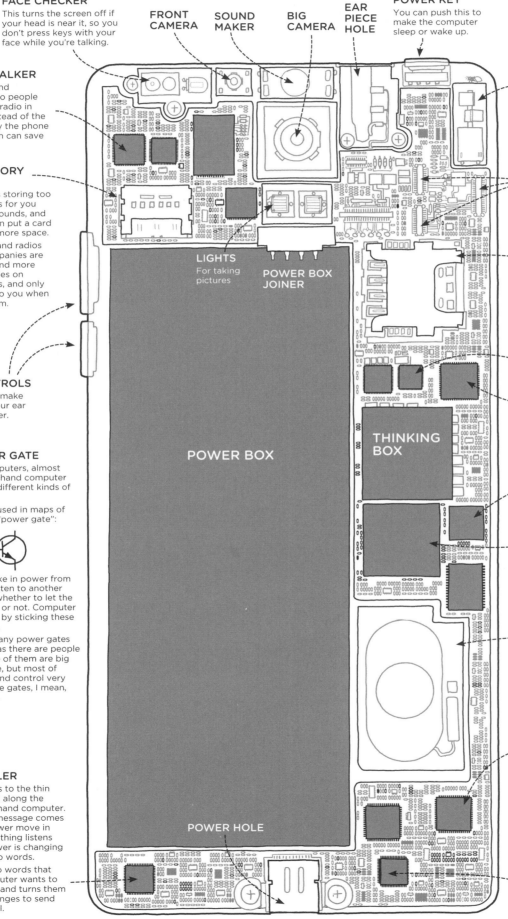

COLORS OF LIGHT

Light is made up of waves, and we see longer waves and shorter waves as different colors. When it rains, the Sun's light hits tiny drops of water and bends as it goes through them. Some colors of light bend more than others, so different colors reach your eye from different parts of the sky.

These sky colors are sorted by how long their waves are, from the shortest waves, blue, to the longest waves, red. But the kinds of light don't stop there! Those are just the shortest and longest waves that our eyes can see.

This picture shows what other colors you would see if the rain colors kept going. (The picture isn't in color, but that's okay—they're not real colors, anyway!)

In real life, even if you could see longer and shorter light waves, you wouldn't see these colors spread out like this in the sky. There are three reasons for this.

First, the Sun gives off most of its light in colors that we can see, and light that's a little shorter or longer. In colors that are *much* shorter or longer, the Sun is pretty dark!

Second, lots of these kinds of light don't go through water, so they wouldn't go through the rain.

Third, the colors are sorted from long to short because short waves (blue) bend more than long waves (red). But there are some colors, kinds we can't see, where it goes the other way! That means that these colors wouldn't be spread out like they're shown here; they'd be laid in sheets over themselves—some parts top to bottom and other parts bottom to top—all in the same area of the sky.

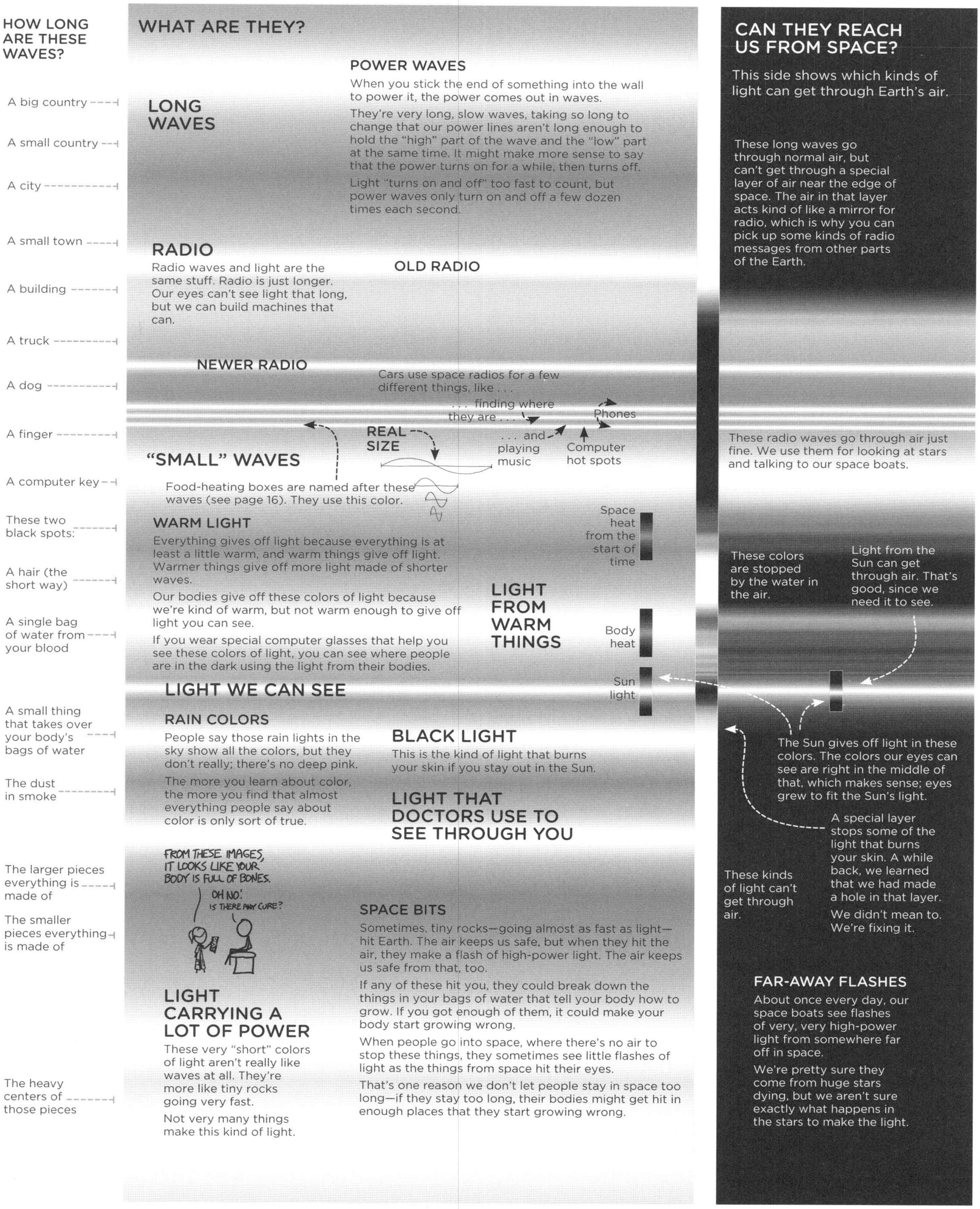

HOW LONG ARE THESE WAVES?

A big country ----
A small country ----
A city -------------
A small town -----
A building ------
A truck ----------
A dog ----------
A finger ---------
A computer key --
These two black spots:
A hair (the short way)
A single bag of water from your blood
A small thing that takes over your body's bags of water
The dust in smoke
The larger pieces everything is made of
The smaller pieces everything is made of
The heavy centers of those pieces

WHAT ARE THEY?

LONG WAVES

POWER WAVES

When you stick the end of something into the wall to power it, the power comes out in waves.

They're very long, slow waves, taking so long to change that our power lines aren't long enough to hold the "high" part of the wave and the "low" part at the same time. It might make more sense to say that the power turns on for a while, then turns off.

Light "turns on and off" too fast to count, but power waves only turn on and off a few dozen times each second.

RADIO

Radio waves and light are the same stuff. Radio is just longer. Our eyes can't see light that long, but we can build machines that can.

OLD RADIO

NEWER RADIO

Cars use space radios for a few different things, like . . .

. . . finding where they are . . .

Phones

. . . and playing music

Computer hot spots

REAL SIZE --

"SMALL" WAVES

Food-heating boxes are named after these waves (see page 16). They use this color.

WARM LIGHT

Everything gives off light because everything is at least a little warm, and warm things give off light. Warmer things give off more light made of shorter waves.

Our bodies give off these colors of light because we're kind of warm, but not warm enough to give off light you can see.

If you wear special computer glasses that help you see these colors of light, you can see where people are in the dark using the light from their bodies.

LIGHT WE CAN SEE

RAIN COLORS

People say those rain lights in the sky show all the colors, but they don't really; there's no deep pink.

The more you learn about color, the more you find that almost everything people say about color is only sort of true.

FROM THESE IMAGES, IT LOOKS LIKE YOUR BODY IS FULL OF BONES.

OH NO! IS THERE ANY CURE?

LIGHT CARRYING A LOT OF POWER

These very "short" colors of light aren't really like waves at all. They're more like tiny rocks going very fast.

Not very many things make this kind of light.

Space heat from the start of time

LIGHT FROM WARM THINGS

Body heat

Sun light

BLACK LIGHT

This is the kind of light that burns your skin if you stay out in the Sun.

LIGHT THAT DOCTORS USE TO SEE THROUGH YOU

SPACE BITS

Sometimes, tiny rocks—going almost as fast as light—hit Earth. The air keeps us safe, but when they hit the air, they make a flash of high-power light. The air keeps us safe from that, too.

If any of these hit you, they could break down the things in your bags of water that tell your body how to grow. If you got enough of them, it could make your body start growing wrong.

When people go into space, where there's no air to stop these things, they sometimes see little flashes of light as the things from space hit their eyes.

That's one reason we don't let people stay in space too long—if they stay too long, their bodies might get hit in enough places that they start growing wrong.

CAN THEY REACH US FROM SPACE?

This side shows which kinds of light can get through Earth's air.

These long waves go through normal air, but can't get through a special layer of air near the edge of space. The air in that layer acts kind of like a mirror for radio, which is why you can pick up some kinds of radio messages from other parts of the Earth.

These radio waves go through air just fine. We use them for looking at stars and talking to our space boats.

These colors are stopped by the water in the air.

Light from the Sun can get through air. That's good, since we need it to see.

The Sun gives off light in these colors. The colors our eyes can see are right in the middle of that, which makes sense; eyes grew to fit the Sun's light.

These kinds of light can't get through air.

A special layer stops some of the light that burns your skin. A while back, we learned that we had made a hole in that layer. We didn't mean to. We're fixing it.

FAR-AWAY FLASHES

About once every day, our space boats see flashes of very, very high-power light from somewhere far off in space.

We're pretty sure they come from huge stars dying, but we aren't sure exactly what happens in the stars to make the light.

THE SKY AT NIGHT

These are some of the things in the sky at night. They're in the sky during the day, too, but the Sun's light makes them too hard to see.

LINES

People like to draw lines between groups of stars to make shapes, then name the group after what they think the shape looks like.

This one is named after a cat.

HAVE THE PEOPLE WHO NAMED THIS EVER *SEEN* A CAT?

(It doesn't look much like a cat to me, either, but names do make star groups easier to remember.)

STARS ARE FAR AWAY

They're so far away that when we look at them, we're seeing how they looked in the past, since their light takes years to reach us.

Sometimes people say that because light takes so long to reach us, the stars we're looking at probably died long ago.

But that's wrong. Most stars you can see are only a few hundred light-years away.

So don't worry; your stars are probably fine!

HOW TO USE A LOOKING GLASS

✓ YES   ✗ NO

TO WAR!

✗ NO   ✗ DEFINITELY NOT

IS THAT A STAR OR A FIRE FLY?

IT'S LANDING ON YOU.

I HOPE IT'S A FIREFLY.

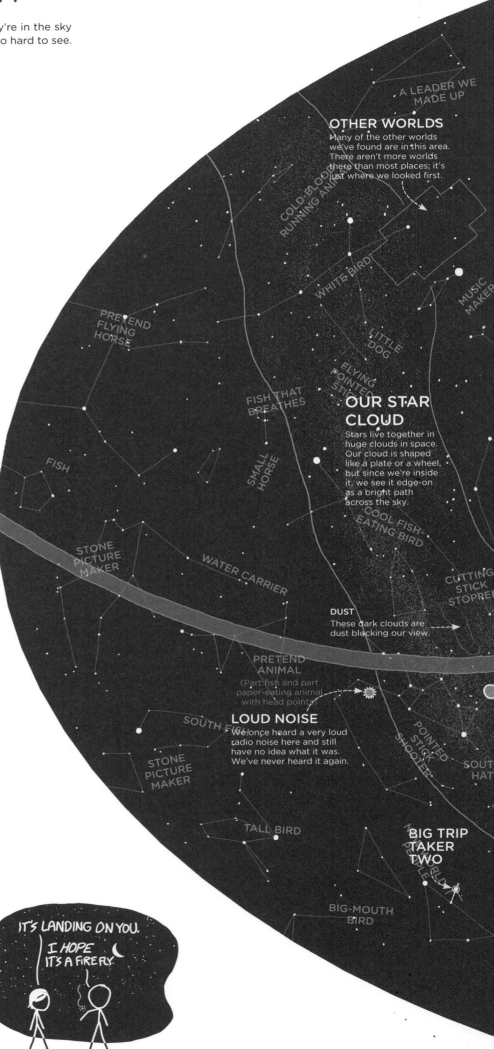

A LEADER WE MADE UP

OTHER WORLDS
Many of the other worlds we've found are in this area. There aren't more worlds there than most places; it's just where we looked first.

COLD-BLOODED RUNNING ANIMAL

WHITE BIRD

MUSIC MAKER

PRETEND FLYING HORSE

LITTLE DOG

FLYING POINTER STICK

FISH THAT BREATHES

OUR STAR CLOUD
Stars live together in huge clouds in space. Our cloud is shaped like a plate or a wheel, but since we're inside it, we see it edge-on as a bright path across the sky.

SMALL HORSE

FISH

COOL FISH-EATING BIRD

CUTTING STICK STOPPER

STONE PICTURE MAKER

WATER CARRIER

DUST
These dark clouds are dust blocking our view.

PRETEND ANIMAL
(Part fish and part paper-eating animal with head points)

SOUTH FISH

LOUD NOISE
We once heard a very loud radio noise here and still have no idea what it was. We've never heard it again.

POINTED STICK SHOOTER

SOUTH HAT

STONE PICTURE MAKER

TALL BIRD

BIG TRIP TAKER TWO

BIG-MOUTH BIRD

OUR STAR

The Sun is a star. It's like other stars, but looks brighter because it's closer. The Sun is so bright that we can only see other stars when its light is blocked by the Earth.

Stars are clouds of air that fell together so hard that they started burning. The Sun's air has been burning since just before the Earth formed, and will keep burning for about that long into the future. After the Sun runs out of air to burn, it will get very big for a short time, blow out lots of heat, then fall together into a small heavy ball that slowly cools.

AIR AROUND THE SUN

The Sun has air around it, like the Earth, but it doesn't have a hard surface under that air. It just keeps getting thicker, all the way to the center.

The air around the Sun is even hotter than some of the inside parts, which is very strange. We're not sure why it's like that.

CENTER

The middle of the Sun is where most of the weight is gathered and where the special fire happens. This special fire only starts if you push air together very, very hard. (This is the fire that powers our biggest city-burning machines.)

FIRE LIGHT

Around the center of the Sun, hot air doesn't rise. Hot air only rises when there's cooler air above it, and near the center of the Sun, *all* the air is hot. Instead, the heat is carried through the Sun by light, just like how light carries the Sun's heat to your face.

The light takes a winding path through the Sun's air. The path is so long that it can take a very long time to reach the surface—as long as hundreds of human lives.

DARK SPOTS

Sometimes dark, cooler spots appear on the Sun, caused by power running through the Sun's surface. Big fire storms often come from places with dark spots.

HOT AIR

The air on fire at the center of the Sun blows light and heat out in all directions. The air in the Sun is trying to fall toward the center, but the light and heat keep blowing it away.

Near the surface of the Sun, the air shakes and rises and turns over, much like a cup of water when you heat it up.

The fire from the center of the Sun heats the air. The air rises and turns over, carrying heat to the surface, where the heat is sent out to space (most of it as light). Some of the air is blown away too, but most of it—cooler, thanks to its trip to space—falls back down to heat up again.

FIRE STORMS

The air in the Sun makes power as it moves (for the same reason that a turning wheel can make power run through metal lines). Sometimes, the power runs through the Sun's surface and blows some of the Sun's fire out into space. These fire storms carry power with them, and if they hit Earth, they can break our computers and power lines.

HOW MUCH HEAT?

Although it's very hot, the fire doesn't actually make new heat very fast. An area of air at the Sun's center makes about as much heat as the body of a cold-blooded animal of the same size.

Even though that doesn't seem like a lot, the Sun is so big—and it has such a thick coat of air around it—that the heat adds up, making it much hotter than any animal.

WHY STARS HAPPEN

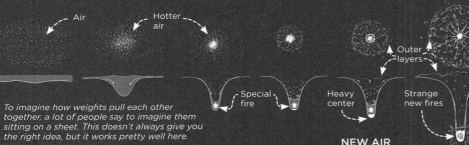

Air — Hotter air

Outer layers — Special fire — Heavy center — Strange new fires

Little white star — Star blown to pieces — Black hole — The last fire

To imagine how weights pull each other together, a lot of people say to imagine them sitting on a sheet. This doesn't always give you the right idea, but it works pretty well here.

AIR CLOUD

A star begins as a cloud of air in space. This cloud is always moving, pushing, and feeling waves go through it, like the surface of the sea.

After a while, a pocket of air happens to get close enough together that the pull of its weight becomes stronger than the force keeping it spread out.

As the air falls together, it gets heavier. This makes it pull harder, which pulls in more air.

As air falls together, it also gets hotter. This heat is how the air pushes back against whatever is pushing it together.

But in this cloud, that heat isn't as strong as the pull of the air's own weight, so it keeps getting smaller and hotter.

SPECIAL FIRE

The air seems like it might keep getting smaller and hotter like this forever. But when it gets hot enough, a new kind of heat is made.

When air is pushed together hard enough, the pieces it's made of can stick together. When they do, they let out a lot of light and heat. This is the heat that powers our largest city-burning war machines.

When a cloud of air gets hot enough, this kind of fire starts, and a great heat blows out from where it burns. This hot wind is strong enough to fight the force pulling the air together. The air gets hotter, but stops getting smaller. A star is born.

The force pushing away and the force pushing in stop each other. If the star falls a little closer together, it makes the fire burn much hotter, pushing it back out.

A star like the Sun has enough air to burn for a very long time—long enough for worlds and life to form. But it can't burn forever.

NEW AIR

When the star burns air by pushing it together, it makes a new kind of heavier air. This kind of air doesn't burn as well, so it gathers, not burned, in the center of the star.

The new air's weight pulls the star together, making the fire burn hotter. The wind from this hotter fire blows the outer parts of the star out farther. Over time, the star grows.

When it starts running out of air to burn, the center falls even closer together, lighting new kinds of fire that blow the outer layers farther away from the star. The star gets very, very big . . . and as the fires begin to die, the force holding back the star's weight disappears, and it starts to fall in on itself.

THE END OF THE EARTH

When the Sun gets very big, its edge will reach the Earth, and Earth will fall in and burn up. You don't need to worry about that now, though. If we want to stay alive past the Sun's death, there are lots of other problems we'll have to face first. Worrying about that one now would be like worrying that one day a tree will grow where you're standing.

THE LAST FIRE

As the dying star falls together, it becomes even hotter than ever before. In this heat, even things that couldn't burn before start to burn, creating new and strange kinds of air. (Much of the stuff we're made of here on Earth comes from a fire like this.)

A lot of heat and light pour out from this last fire, and for a moment, the star can become the brightest thing in all of space.

WHAT'S LEFT

The heat blows much of the star away into space. Sometimes, what's left of the star will fall together until it becomes a bright white ball of hard air that slowly cools. Someday, this will happen to the Sun.

If the star is bigger than the Sun, it may have too much weight to stop even there. The weight of the hard ball will make it keep falling in on itself, until it becomes so strong it pulls in even light, leaving behind a black hole in space.

HOW TO COUNT THINGS

To count how heavy things are, we pick a weight to call "one." Then, if you say a weight is "ten," people understand it's as heavy as ten "ones." We do the same for counting other things, like how fast or hot things are.

People don't always agree on how much "one" is, which can cause a lot of problems. A space boat once missed a world because people got confused about which "one" they should be using for weight.

Most countries have agreed to make "one" the same thing everywhere. Here's what numbers from one to ten hundred mean in that counting system.

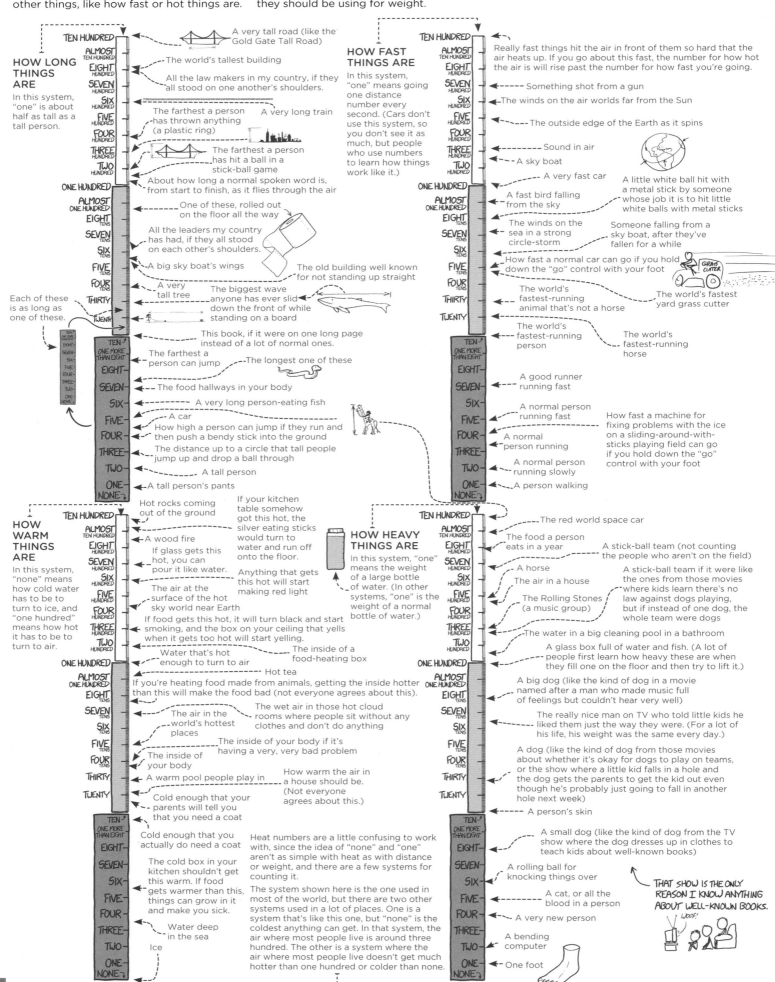

HOW LONG THINGS ARE

In this system, "one" is about half as tall as a tall person.

TEN HUNDRED
ALMOST TEN HUNDRED
EIGHT HUNDRED
SEVEN HUNDRED
SIX HUNDRED
FIVE HUNDRED
FOUR HUNDRED
THREE HUNDRED
TWO HUNDRED
ONE HUNDRED
ALMOST ONE HUNDRED
EIGHT TENS
SEVEN TENS
SIX TENS
FIVE TENS
FOUR TENS
THIRTY
TWENTY
TEN / ONE MORE THAN EIGHT
EIGHT
SEVEN
SIX
FIVE
FOUR
THREE
TWO
ONE
NONE

- A very tall road (like the Gold Gate Tall Road)
- The world's tallest building
- All the law makers in my country, if they all stood on one another's shoulders.
- The farthest a person has thrown anything (a plastic ring)
- A very long train
- The farthest a person has hit a ball in a stick-ball game
- About how long a normal spoken word is, from start to finish, as it flies through the air
- One of these, rolled out on the floor all the way
- All the leaders my country has had, if they all stood on each other's shoulders.
- A big sky boat's wings
- The old building well known for not standing up straight
- A very tall tree
- The biggest wave anyone has ever slid down the front of while standing on a board
- This book, if it were on one long page instead of a lot of normal ones.
- The farthest a person can jump
- The longest one of these
- The food hallways in your body
- A very long person-eating fish
- A car
- How high a person can jump if they run and then push a bendy stick into the ground
- The distance up to a circle that tall people jump up and drop a ball through
- A tall person
- A tall person's pants

Each of these is as long as one of these.

HOW FAST THINGS ARE

In this system, "one" means going one distance number every second. (Cars don't use this system, so you don't see it as much, but people who use numbers to learn how things work like it.)

Really fast things hit the air in front of them so hard that the air heats up. If you go about this fast, the number for how hot the air is will rise past the number for how fast you're going.

- Something shot from a gun
- The winds on the air worlds far from the Sun
- The outside edge of the Earth as it spins
- Sound in air
- A sky boat
- A very fast car
- A fast bird falling from the sky
- A little white ball hit with a metal stick by someone whose job it is to hit little white balls with metal sticks
- The winds on the sea in a strong circle-storm
- Someone falling from a sky boat, after they've fallen for a while
- How fast a normal car can go if you hold down the "go" control with your foot
- The world's fastest-running animal that's not a horse
- The world's fastest yard grass cutter
- The world's fastest-running person
- The world's fastest-running horse
- A good runner running fast
- A normal person running fast
- How fast a machine for fixing problems with the ice on a sliding-around-with-sticks playing field can go if you hold down the "go" control with your foot
- A normal person running
- A normal person running slowly
- A person walking

HOW WARM THINGS ARE

In this system, "none" means how cold water has to be to turn to ice, and "one hundred" means how hot it has to be to turn to air.

- Hot rocks coming out of the ground
- A wood fire
- If glass gets this hot, you can pour it like water.
- If your kitchen table somehow got this hot, the silver eating sticks would turn to water and run off onto the floor.
- Anything that gets this hot will start making red light
- The air at the surface of the hot sky world near Earth
- If food gets this hot, it will turn black and start smoking, and the box on your ceiling that yells when it gets too hot will start yelling.
- Water that's hot enough to turn to air
- The inside of a food-heating box
- Hot tea
- If you're heating food made from animals, getting the inside hotter than this will make the food bad (not everyone agrees about this).
- The wet air in those hot cloud rooms where people sit without any clothes and don't do anything
- The air in the world's hottest places
- The inside of your body if it's having a very, very bad problem
- The inside of your body
- A warm pool people play in
- How warm the air in a house should be. (Not everyone agrees about this.)
- Cold enough that your parents will tell you that you need a coat
- Cold enough that you actually do need a coat
- The cold box in your kitchen shouldn't get this warm. If food gets warmer than this, things can grow in it and make you sick.
- Water deep in the sea
- Ice

Heat numbers are a little confusing to work with, since the idea of "none" and "one" aren't as simple with heat as with distance or weight, and there are a few systems for counting it.

The system shown here is the one used in most of the world, but there are two other systems used in a lot of places. One is a system that's like this one, but "none" is the coldest anything can get. In that system, the air where most people live is around three hundred. The other is a system where the air where most people live doesn't get much hotter than one hundred or colder than none.

HOW HEAVY THINGS ARE

In this system, "one" means the weight of a large bottle of water. (In other systems, "one" is the weight of a normal bottle of water.)

- The red world space car
- The food a person eats in a year
- A stick-ball team (not counting the people who aren't on the field)
- A horse
- A stick-ball team if it were like the ones from those movies where kids learn there's no law against dogs playing, but if instead of one dog, the whole team were dogs
- The air in a house
- The Rolling Stones (a music group)
- The water in a big cleaning pool in a bathroom
- A glass box full of water and fish. (A lot of people first learn how heavy these are when they fill one on the floor and then try to lift it.)
- A big dog (like the kind of dog in a movie named after a man who made music full of feelings but couldn't hear very well)
- The really nice man on TV who told little kids he liked them just the way they were. (For a lot of his life, his weight was the same every day.)
- A dog (like the kind of dog from those movies about whether it's okay for dogs to play on teams, or the show where a little kid falls in a hole and the dog gets the parents to get the kid out even though he's probably just going to fall in another hole next week)
- A person's skin
- A small dog (like the kind of dog from the TV show where the dog dresses up in clothes to teach kids about well-known books)
- A rolling ball for knocking things over
- A cat, or all the blood in a person
- A very new person
- A bending computer
- One foot

THAT SHOW IS THE ONLY REASON I KNOW ANYTHING ABOUT WELL-KNOWN BOOKS.
WOOF!

ROOM FOR HELPING PEOPLE

Little things go wrong in our bodies all the time, but bodies are pretty good at fixing problems. Parts are always breaking, but our bodies make new ones. Tiny living things try to make us sick, but our bodies are full of groups of tiny machines flying all over the place looking for things that don't belong and getting rid of them. Usually, these problems get fixed without us even knowing!

But just as there are places we can't travel without help from other people and machines—like the Moon, or the bottom of the sea—there are problems we can't fix without help from other people and machines.

When we're sick, or something goes wrong with us, sometimes we need to visit a room full of machines like these so we can talk to doctors and get the help we need.

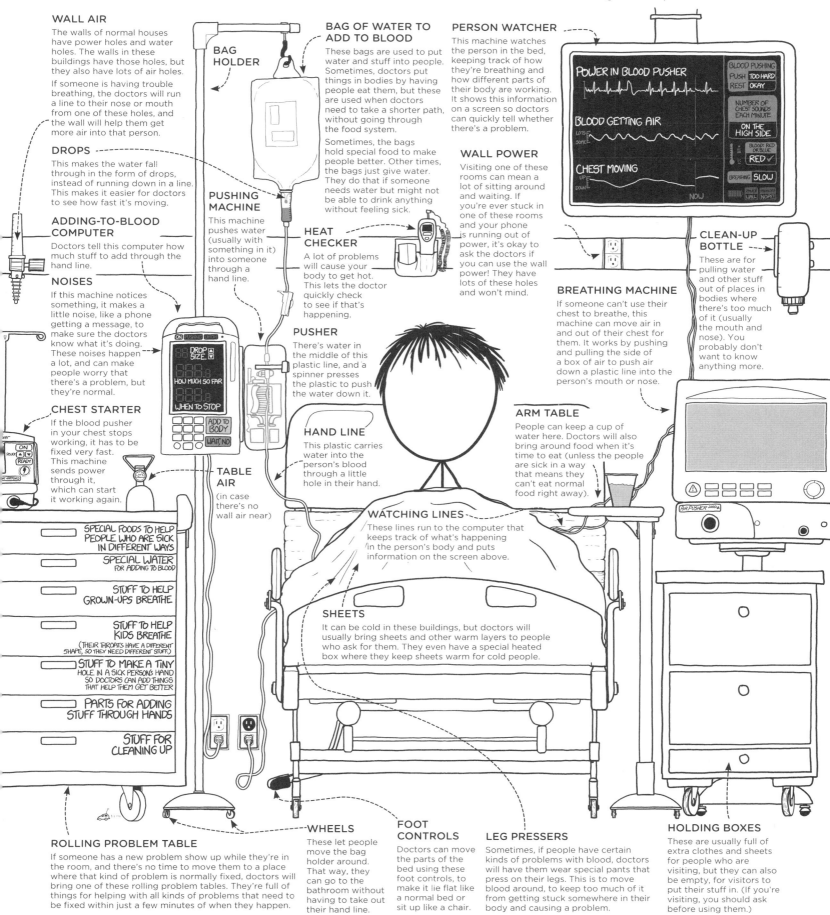

WALL AIR

The walls of normal houses have power holes and water holes. The walls in these buildings have those holes, but they also have lots of air holes.

If someone is having trouble breathing, the doctors can run a line to their nose or mouth from one of these holes, and the wall will help them get more air into that person.

DROPS

This makes the water fall through in the form of drops, instead of running down in a line. This makes it easier for doctors to see how fast it's moving.

ADDING-TO-BLOOD COMPUTER

Doctors tell this computer how much stuff to add through the hand line.

NOISES

If this machine notices something, it makes a little noise, like a phone getting a message, to make sure the doctors know what it's doing. These noises happen a lot, and can make people worry that there's a problem, but they're normal.

CHEST STARTER

If the blood pusher in your chest stops working, it has to be fixed very fast. This machine sends power through it, which can start it working again.

BAG HOLDER

PUSHING MACHINE

This machine pushes water (usually with something in it) into someone through a hand line.

BAG OF WATER TO ADD TO BLOOD

These bags are used to put water and stuff into people. Sometimes, doctors put things in bodies by having people eat them, but these are used when doctors need to take a shorter path, without going through the food system.

Sometimes, the bags hold special food to make people better. Other times, the bags just give water. They do that if someone needs water but might not be able to drink anything without feeling sick.

HEAT CHECKER

A lot of problems will cause your body to get hot. This lets the doctor quickly check to see if that's happening.

PUSHER

There's water in the middle of this plastic line, and a spinner presses the plastic to push the water down it.

HAND LINE

This plastic carries water into the person's blood through a little hole in their hand.

TABLE AIR

(in case there's no wall air near)

PERSON WATCHER

This machine watches the person in the bed, keeping track of how they're breathing and how different parts of their body are working. It shows this information on a screen so doctors can quickly tell whether there's a problem.

WALL POWER

Visiting one of these rooms can mean a lot of sitting around and waiting. If you're ever stuck in one of these rooms and your phone is running out of power, it's okay to ask the doctors if you can use the wall power! They have lots of these holes and won't mind.

POWER IN BLOOD PUSHER

BLOOD GETTING AIR

LOTS
SOME

CHEST MOVING

UP
DOWN

NOW

BLOOD PUSHING
PUSH TOO HARD
REST OKAY

NUMBER OF CHEST SOUNDS EACH MINUTE

ON THE HIGH SIDE

BLOOD: RED OR BLUE
RED ✓

BREATHING SLOW

BREATHING MACHINE

If someone can't use their chest to breathe, this machine can move air in and out of their chest for them. It works by pushing and pulling the side of a box of air to push air down a plastic line into the person's mouth or nose.

CLEAN-UP BOTTLE

These are for pulling water and other stuff out of places in bodies where there's too much of it (usually the mouth and nose). You probably don't want to know anything more.

ARM TABLE

People can keep a cup of water here. Doctors will also bring around food when it's time to eat (unless the people are sick in a way that means they can't eat normal food right away).

WATCHING LINES

These lines run to the computer that keeps track of what's happening in the person's body and puts information on the screen above.

SHEETS

It can be cold in these buildings, but doctors will usually bring sheets and other warm layers to people who ask for them. They even have a special heated box where they keep sheets warm for cold people.

AIR PUSHER 2100A

SPECIAL FOODS TO HELP PEOPLE WHO ARE SICK IN DIFFERENT WAYS

SPECIAL WATER FOR ADDING TO BLOOD

STUFF TO HELP GROWN-UPS BREATHE

STUFF TO HELP KIDS BREATHE (THEIR THROATS HAVE A DIFFERENT SHAPE, SO THEY NEED DIFFERENT STUFF.)

STUFF TO MAKE A TINY HOLE IN A SICK PERSON'S HAND SO DOCTORS CAN ADD THINGS THAT HELP THEM GET BETTER

PARTS FOR ADDING STUFF THROUGH HANDS

STUFF FOR CLEANING UP

ROLLING PROBLEM TABLE

If someone has a new problem show up while they're in the room, and there's no time to move them to a place where that kind of problem is normally fixed, doctors will bring one of these rolling problem tables. They're full of things for helping with all kinds of problems that need to be fixed within just a few minutes of when they happen.

WHEELS

These let people move the bag holder around. That way, they can go to the bathroom without having to take out their hand line.

FOOT CONTROLS

Doctors can move the parts of the bed using these foot controls, to make it lie flat like a normal bed or sit up like a chair.

LEG PRESSERS

Sometimes, if people have certain kinds of problems with blood, doctors will have them wear special pants that press on their legs. This is to move blood around, to keep too much of it from getting stuck somewhere in their body and causing a problem.

HOLDING BOXES

These are usually full of extra clothes and sheets for people who are visiting, but they can also be empty, for visitors to put their stuff in. (If you're visiting, you should ask before using them.)

PLAYING FIELDS

And how big they are.

(The real-life fields are all ten hundred times larger than these drawings.)

A lot of games are played by throwing things, kicking things, hitting things, and using sticks. Games put these together in different ways:

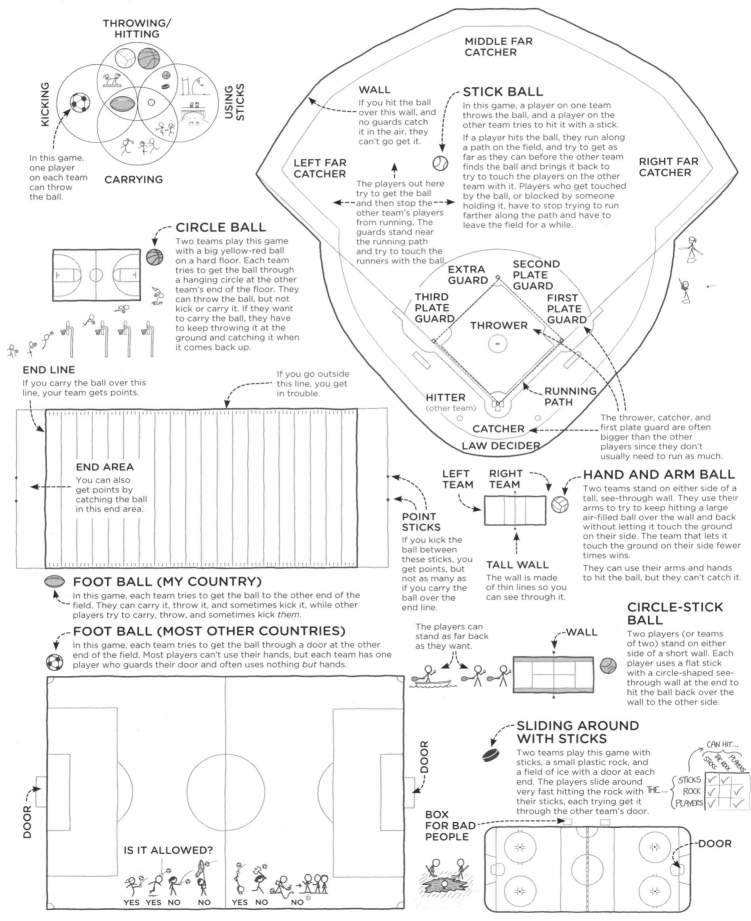

THROWING/ HITTING

KICKING

USING STICKS

CARRYING

In this game, one player on each team can throw the ball.

STICK BALL

In this game, a player on one team throws the ball, and a player on the other team tries to hit it with a stick.

If a player hits the ball, they run along a path on the field, and try to get as far as they can before the other team finds the ball and brings it back to try to touch the players on the other team with it. Players who get touched by the ball, or blocked by someone holding it, have to stop trying to run farther along the path and have to leave the field for a while.

MIDDLE FAR CATCHER

WALL
If you hit the ball over this wall, and no guards catch it in the air, they can't go get it.

LEFT FAR CATCHER

RIGHT FAR CATCHER

The players out here try to get the ball and then stop the other team's players from running. The guards stand near the running path and try to touch the runners with the ball.

EXTRA GUARD

SECOND PLATE GUARD

THIRD PLATE GUARD

FIRST PLATE GUARD

THROWER

HITTER (other team)

RUNNING PATH

CATCHER

LAW DECIDER

The thrower, catcher, and first plate guard are often bigger than the other players since they don't usually need to run as much.

CIRCLE BALL

Two teams play this game with a big yellow-red ball on a hard floor. Each team tries to get the ball through a hanging circle at the other team's end of the floor. They can throw the ball, but not kick or carry it. If they want to carry the ball, they have to keep throwing it at the ground and catching it when it comes back up.

HELP!

END LINE
If you carry the ball over this line, your team gets points.

If you go outside this line, you get in trouble.

END AREA
You can also get points by catching the ball in this end area.

FOOT BALL (MY COUNTRY)
In this game, each team tries to get the ball to the other end of the field. They can carry it, throw it, and sometimes kick it, while other players try to carry, throw, and sometimes kick *them*.

FOOT BALL (MOST OTHER COUNTRIES)
In this game, each team tries to get the ball through a door at the other end of the field. Most players can't use their hands, but each team has one player who guards their door and often uses nothing *but* hands.

DOOR

DOOR

IS IT ALLOWED?

YES YES NO NO   YES NO NO

HAND AND ARM BALL
Two teams stand on either side of a tall, see-through wall. They use their arms to try to keep hitting a large air-filled ball over the wall and back without letting it touch the ground on their side. The team that lets it touch the ground on their side fewer times wins.

They can use their arms and hands to hit the ball, but they can't catch it.

LEFT TEAM

RIGHT TEAM

POINT STICKS
If you kick the ball between these sticks, you get points, but not as many as if you carry the ball over the end line.

TALL WALL
The wall is made of thin lines so you can see through it.

The players can stand as far back as they want.

CIRCLE-STICK BALL
Two players (or teams of two) stand on either side of a short wall. Each player uses a flat stick with a circle-shaped see-through wall at the end to hit the ball back over the wall to the other side.

WALL

SLIDING AROUND WITH STICKS
Two teams play this game with sticks, a small plastic rock, and a field of ice with a door at each end. The players slide around very fast hitting the rock with their sticks, each trying get it through the other team's door.

CAN HIT...

	THE ROCK	PLAYERS
STICKS	✓	✓
ROCK	✓	
PLAYERS	✓		✓

BOX FOR BAD PEOPLE

DOOR

EARTH'S PAST
Everything* that has happened here so far *NOT QUITE EVERYTHING

We learn about the history of the Earth from rocks. Rocks are laid down in layers, and by looking at the layers from different parts of the world, which are all different ages, we can piece together a single history that goes back almost to the start of the world.

This picture shows what it would look like if you could see the whole history of Earth in a single set of layers, with every year as thick as every other. In real life, no single place has all these layers together, and there are no layers at all from the oldest part of the Earth's history.

All of human history, since we first learned to write and build cities, is a layer as thin as a piece of paper.

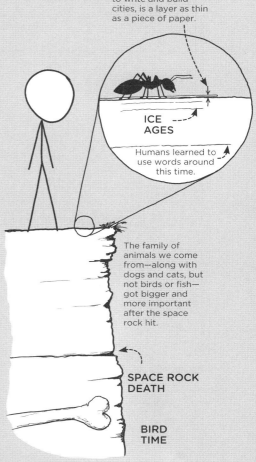

ICE AGES

Humans learned to use words around this time.

NOW

CAREFUL!

SPACE ROCK HITS EARTH
A big rock hit the Earth, and lots of the animals died. Some groups lived, like birds, some kinds of fish, and our parents.

THE BIRD TIME

THE BIRD TIME
A big, well-known group of animals lived during this time. Today's birds are the only animals from that family alive now, but many other animals came from it in the past—like big ones with long necks and bitey ones with huge teeth.

THE TREE TIME

THE GREAT DYING
Almost everything died here, and we're not sure why. There were lots of strange changes in the air and the sea, and around that time a huge layer of hot rock came up out of the Earth and covered a large part of the land. So whatever happened, it was pretty bad.

"The Great Dying" sounds like a name made up to use simple words, but it's not; serious people call it that.

The family of animals we come from—along with dogs and cats, but not birds or fish—got bigger and more important after the space rock hit.

EVERYTHING GETS COLD
The Earth got really cold here, and ice covered lots of it, even parts around the middle where it's usually hot.

LIFE GETS BIG AND STRANGE
Around this time, big animals started to appear. If you find rocks from this time, you can see lots of strange things in them.

LAND COMES TOGETHER AND BREAKS UP
Right now, Earth's land is broken up into five or six big areas with water in between, but before that, it was pushed together. We think this breaking up and pushing together happened a few times, although it's hard to tell how many.

THE SIMPLE TIME
For a long time, life was pretty simple. There were no animals. Most life was small, either made of single bags of water moving around alone, or big groups of bags growing in big piles on the sea floor.

THE GREAT AIR CHANGE
Around this time, the air changed. A kind of life appeared that ate the Sun's light and breathed out a new kind of air. This new air probably killed almost everything else, and for the first time it made fire possible. But it's also the part of air we need to breathe, so it was good for us!

Trees and flowers do the same kind of breathing as that early life. We think the things in their leaves that let them eat the Sun's light—the things which make them green—are the children of the life that changed the air.

SPACE ROCK DEATH

BIRD TIME

SPACE ROCK HITS EARTH

SPACE ROCK HITS EARTH

RED METAL LINES
There was once a kind of metal that was spread out in all the waters of the sea (the same way the white stuff we put on food is now).

When the air changed, the water changed too. The metal turned red and fell to the bottom of the sea. It left beautiful red lines in the rocks.

We use the metal from those layers to make things like machines and buildings.

THE GREAT ROCK FALL
Most of the big circles on the Moon seem to be from around this time, which makes us think there were a lot of rocks flying around hitting worlds around then.

The rocks might have been thrown at us by the big air worlds far from the Sun. As they settled into their circle paths—some of them may have changed places!—their pull would have changed the path of the rocks around them, and some of those might have hit us.

If the rocks hit the Moon, they probably hit the Earth (and other worlds near us) too, and might have made the land run like water and the seas turn to air.

FIRST SIGNS OF LIFE
The first signs of life appear in these rocks. We've found some black rocks (the kind used in writing sticks) that we think must have come from living things.

But there are very few rocks from this time, and they're old and hard to understand for sure.

OLDER LIFE?
All life is part of one family, and the information stored in our water bags changes over time, as animals have children and those children have children. By looking at the information stored in the water bags of living things, doctors can figure out how long ago their shared parents lived.

When people have tried to work out how old life's shared parent is, they sometimes come up with a number that's a little *older* than the great rock fall.

But we think the seas turned to air and the rocks to fire, and it's hard to understand how anything could have lived through that.

EARTH FORMED
The Earth formed from the same cloud that the Sun and other worlds did, at around the same time. It was hot when it formed, but we think it must have cooled off pretty quickly, because we've seen signs that there was water almost right away.

QUESTION TIME
This picture shows rock layers back to the start of the Earth, but in real life, there aren't any big areas of rock left over from before this time, so it's hard to say what it was like. We think there were seas, at least for part of it, but we're not sure what it was like.

MOON FORMED
We think the Earth got hit by another world here, while it was forming, and all the rock that got thrown free turned into the Moon.

TREE OF LIFE

All life (that we know of) is part of a family. We all come from one living thing that appeared in the early days of the Earth. That living thing grew, had children, and changed over time. People, trees, grass, and flowers are all children of that first life.

As living things make more living things, the information they pass to them changes, making the new things a little different from the old. Over time, these small changes can lead to very different kinds of living things growing from one. This tree shows how different kinds of life branched off from one another.

This tree doesn't show all living things, or even most of them. It just shows some of the living things you might know, along with which branch of life's family they're in.

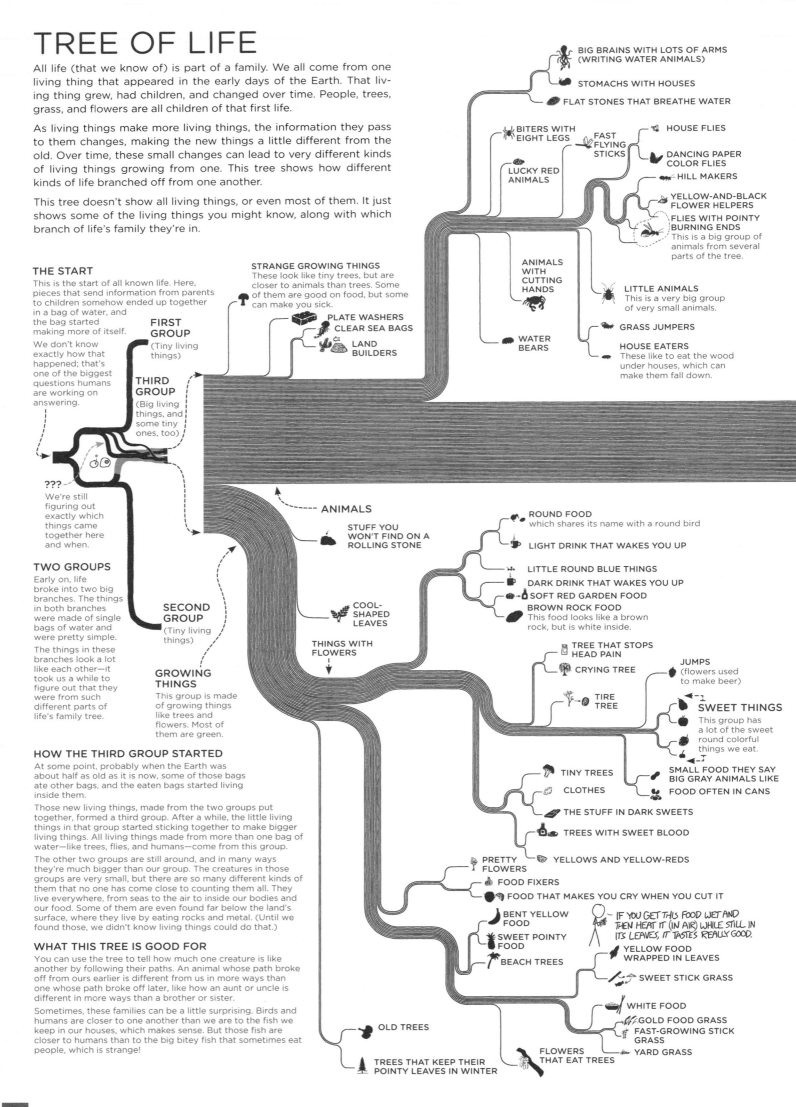

THE START
This is the start of all known life. Here, pieces that send information from parents to children somehow ended up together in a bag of water, and the bag started making more of itself.
We don't know exactly how that happened; that's one of the biggest questions humans are working on answering.

???
We're still figuring out exactly which things came together here and when.

TWO GROUPS
Early on, life broke into two big branches. The things in both branches were made of single bags of water and were pretty simple.
The things in these branches look a lot like each other—it took us a while to figure out that they were from such different parts of life's family tree.

HOW THE THIRD GROUP STARTED
At some point, probably when the Earth was about half as old as it is now, some of those bags ate other bags, and the eaten bags started living inside them.
Those new living things, made from the two groups put together, formed a third group. After a while, the little living things in that group started sticking together to make bigger living things. All living things made from more than one bag of water—like trees, flies, and humans—come from this group.
The other two groups are still around, and in many ways they're much bigger than our group. The creatures in those groups are very small, but there are so many different kinds of them that no one has come close to counting them all. They live everywhere, from seas to the air to inside our bodies and our food. Some of them are even found far below the land's surface, where they live by eating rocks and metal. (Until we found those, we didn't know living things could do that.)

WHAT THIS TREE IS GOOD FOR
You can use the tree to tell how much one creature is like another by following their paths. An animal whose path broke off from ours earlier is different from us in more ways than one whose path broke off later, like how an aunt or uncle is different in more ways than a brother or sister.
Sometimes, these families can be a little surprising. Birds and humans are closer to one another than we are to the fish we keep in our houses, which makes sense. But those fish are closer to humans than to the big bitey fish that sometimes eat people, which is strange!

FIRST GROUP
(Tiny living things)

THIRD GROUP
(Big living things, and some tiny ones, too)

STRANGE GROWING THINGS
These look like tiny trees, but are closer to animals than trees. Some of them are good on food, but some can make you sick.

PLATE WASHERS
CLEAR SEA BAGS
LAND BUILDERS

SECOND GROUP
(Tiny living things)

GROWING THINGS
This group is made of growing things like trees and flowers. Most of them are green.

ANIMALS
STUFF YOU WON'T FIND ON A ROLLING STONE

COOL-SHAPED LEAVES

THINGS WITH FLOWERS

BIG BRAINS WITH LOTS OF ARMS (WRITING WATER ANIMALS)
STOMACHS WITH HOUSES
FLAT STONES THAT BREATHE WATER

BITERS WITH EIGHT LEGS
FAST FLYING STICKS
LUCKY RED ANIMALS

HOUSE FLIES
DANCING PAPER COLOR FLIES
HILL MAKERS
YELLOW-AND-BLACK FLOWER HELPERS
FLIES WITH POINTY BURNING ENDS
This is a big group of animals from several parts of the tree.

ANIMALS WITH CUTTING HANDS

LITTLE ANIMALS
This is a very big group of very small animals.

GRASS JUMPERS

WATER BEARS

HOUSE EATERS
These like to eat the wood under houses, which can make them fall down.

ROUND FOOD
which shares its name with a round bird
LIGHT DRINK THAT WAKES YOU UP
LITTLE ROUND BLUE THINGS
DARK DRINK THAT WAKES YOU UP
SOFT RED GARDEN FOOD
BROWN ROCK FOOD
This food looks like a brown rock, but is white inside.

TREE THAT STOPS HEAD PAIN
CRYING TREE
TIRE TREE

JUMPS
(flowers used to make beer)

SWEET THINGS
This group has a lot of the sweet round colorful things we eat.

TINY TREES
CLOTHES
THE STUFF IN DARK SWEETS
TREES WITH SWEET BLOOD

SMALL FOOD THEY SAY BIG GRAY ANIMALS LIKE
FOOD OFTEN IN CANS

YELLOWS AND YELLOW-REDS

PRETTY FLOWERS
FOOD FIXERS
FOOD THAT MAKES YOU CRY WHEN YOU CUT IT

BENT YELLOW FOOD
SWEET POINTY FOOD
BEACH TREES

IF YOU GET THIS FOOD WET AND THEN HEAT IT (IN AIR) WHILE STILL IN ITS LEAVES, IT TASTES REALLY GOOD.

YELLOW FOOD WRAPPED IN LEAVES
SWEET STICK GRASS
WHITE FOOD
GOLD FOOD GRASS
FAST-GROWING STICK GRASS
YARD GRASS

OLD TREES

FLOWERS THAT EAT TREES

TREES THAT KEEP THEIR POINTY LEAVES IN WINTER

ANIMALS WITH BONES

The animals in this part of the tree have bones inside them. Some of the animals in other parts of the tree have hard body parts, but they usually have them on the outside. The animals in this part have bones on the inside, with the soft parts hanging from them.

BIG BITEY FISH
These eat people sometimes, but not very often.

NORMAL FISH
(Some of these bite, too.)

WATER JUMPERS

WATER JUMPERS WITH LONG BACK ENDS
These are like water jumpers, but have long back ends and don't jump.

ANIMALS WITH HAIR
We're part of this group. These animals usually have hair, make white water for babies to drink, and don't lay eggs.

STRANGE ANIMAL
This animal looks like it's part cat, part fish, and part bird. It broke off from the other hair animals early, so it's very different and strange.

PRETEND SEA LADIES
These look nothing like sea ladies, but people used to pretend they were.

BIG GRAY ARM-NOSES

SLOW CLIMBERS

ROLL-UP CATS WITH HARD SKIN

ANIMAL EATERS
These animals mostly eat other animals. There are two main kinds: cat-shaped and dog-shaped.

(Cats and dogs are in those groups, of course, but other animals like bears are too.)

CAT-SHAPED

SKIN BIRDS
People think these are close to the little house-food eaters with big teeth, but they're actually more like huge air fish and horses.

← DOG-SHAPED

LONG BITEY DOGS
SMELLY DOGS
RIVER DOGS
SEA DOGS
BEARS
—THE DOG FAMILY
DOGS (NOT OUR FRIENDS)
DOGS (OUR FRIENDS)
SMALL DOGS
TINY SCREAMING DOGS

DOG-SHAPED

CAT WITH LINES
SPOTTED CAT (OLD WORLD)
GREAT CAT
SPOTTED CAT (NEW WORLD)
SNOW CAT
HOUSE CAT
FAST CAT
MOUNTAIN CAT
This cat has many names. A lot of people don't know they're all names for the same animal.

CAT-SHAPED

LAUGHING CAT

THE CAT FAMILY

ALMOST CAT
This animal looks like a cat with a long neck. It's the closest thing to a cat that's not actually in the cat family.

I WANT ONE!
MROUL?

PINK ANIMAL WE EAT
LONG NECK
BIG FOOD ANIMAL
RUNNER IN THE TREES
ANGRY RIVER ANIMAL
AIR-BREATHING FISH (NOT FISH)

SAND HORSE
HORSE
STORE CHECK-OUT HORSE
GRAY TRUCK ANIMAL WITH A POINTY FACE

BODY HEAT
Some animals in this group get most of their heat from the world around them instead of from their bodies. When the world gets cold, they do too.

Not every animal in this family is like that. A few of them, like birds, keep themselves warm the way we do.

POCKET BABIES
Many of these animals keep their babies in pockets and feed them there.

BABY-FEEDING BAGS
These animals are joined to their babies with a feeding bag until the baby is born.

SLOW NIGHT WALKERS

FACE-BITING DOGS
JUMPERS WITH POCKETS

COLD BLOOD WALL WALKERS
LONG BITERS WITH NO ARMS OR LEGS

SLOW ROCKS WITH LEGS AND A HEAD

ANIMAL THAT LOOKS LIKE A TREE IN THE WATER
. . . but it can eat you.

ANIMALS WITH BIG FRONT TEETH

LITTLE HOUSE-FOOD EATERS
RIVER STOPPERS
GRAY TREE-JUMPERS
POINTY CATS
JUMPERS WITH LONG EARS

THE FAMILY BIRDS ARE FROM
Birds are the living members of a very well-known family. Some of the animals from that family were the biggest land animals that ever lived.

They lived, grew, and changed for a very long time. When a space rock hit Earth, most of the ones that were alive at that time died out, but some groups didn't. We call the branch that those groups are from "birds."

Sometimes, you'll hear people say that birds came from that family, but that they're not really *part* of it. This is wrong! Just about any way you count it, birds are part of that family.

x **THE POINTY KIND**
x **THE KIND WITH PLATES**
x **THE BITEY KIND**
✓ **BIRDS**
x **THE LONG KIND**

HAND ANIMALS
These animals are good at climbing. We're in this group.

BIG HAND-WALKERS
FRIENDLY HAND ANIMALS
HAND ANIMALS THAT USE STICKS
HUMANS
STRONG ARMS

HUMAN-SHAPED

TINY CLIMBERS
Some of the animals in this group are smaller than your hand!

This is just a tiny part of the tree of life. The whole tree is too big to fit in any single picture, and there are too many kinds of life for anyone to give names to all of them—no matter what kind of words they use.

And really, a true tree of life wouldn't just have a line for every *kind* of life. It would have a line for each living thing that ever was, every one of them crossing and joining and winding across the page, slowly changing from one kind of life to another, in a path that reaches all the way back, without a single break, to that very first life.

No one really knows how many living things there are in the world, but we can make some guesses, and they're big. Not only can we never find enough words to talk about all those lives, we have a hard time talking about the number itself.

Here's one way to think about how many things have lived on Earth: The world is covered in seas that are ringed with beaches of sand. One day, when you're walking on a beach, pick up some sand and look at it. Imagine that every tiny piece of sand under your feet is a whole world of its own, each one with its own seas and beaches, just like Earth.

The full tree of life has as many living things as there are bits of sand on all those beaches on all those tiny sand worlds put together.

Next to the world we're talking about, all our words are small.

THE TEN HUNDRED WORDS PEOPLE USE THE MOST

This is my set of the ten hundred words people use the most.

There are lots of different ways of counting how much people use a word. You can look at what words people use in TV shows, in books, in news stories, in the letters they write, or when sending computer messages. You can also look at words that are most used right now, or at words that have been used for the past ten years, or the past hundred. You can look at all books, or books of made-up stories, books of history, or well-known old books. These different ways of counting all come up with different sets of words people use the most.

I wanted to write this book using words that sounded familiar and simple. To choose the set of ten hundred words I would use, I looked at many sets of words put together in different ways (I even made one by counting the words in computer messages people had sent me). I especially looked at sets of words made from books that told made-up stories, since I found that counting how much a word was used in those books fit well with how "simple" it sounded.

If the different sets agreed that a word was used a lot, I added it to my ten hundred. If they didn't agree on a word, I used my sense of how simple the word was to decide whether it should be in the ten hundred.

Here are the words I chose. If you want to try explaining something using only these ten hundred words, you can use xkcd.com/simplewriter to check your words as you write!

a	anywhere	below	burn	climb
able	apartment	bend	bus	close
about	appear	beneath	business	clothes
above	approach	beside	busy	cloud
accept	area	best	but	coat
across	arm	better	buy	coffee
act	around	between	by	cold
actually	arrive	beyond	call	college
add	art	big	calm	color
admit	as	bird	camera	come
afraid	ask	bit	can	company
after	asleep	bite	car	completely
afternoon	at	black	card	computer
again	attack	block	care	confuse
against	attention	blood	careful	consider
age	aunt	blow	carefully	continue
ago	avoid	blue	carry	control
agree	away	board	case	conversation
ahead	baby	boat	cat	cool
air	back	body	catch	cop
alive	bad	bone	cause	corner
all	bag	book	ceiling	count
allow	ball	boot	center	counter
almost	bank	bore	certain	country
alone	bar	both	certainly	couple
along	barely	bother	chair	course
already	bathroom	bottle	chance	cover
also	be	bottom	change	crazy
although	beach	box	check	create
always	bear	boy	cheek	creature
among	beat	brain	chest	cross
and	beautiful	branch	child	crowd
angry	because	break	choice	cry
animal	become	breast	choose	cup
another	bed	breath	church	cut
answer	bedroom	breathe	cigarette	dad
any	beer	bridge	circle	dance
anybody	before	bright	city	dark
anymore	begin	bring	class	darkness
anyone	behind	brother	clean	daughter
anything	believe	brown	clear	day
anyway	belong	building	clearly	dead

death	except	funny	history	law
decide	excite	future	hit	lay
deep	expect	game	hold	lead
desk	explain	garden	hole	leaf
despite	expression	gate	home	lean
die	extra	gather	hope	learn
different	eye	gently	horse	leave
dinner	face	get	hospital	leg
direction	fact	gift	hot	less
dirt	fade	girl	hotel	let
disappear	fail	give	hour	letter
discover	fall	glance	house	lie
distance	familiar	glass	how	life
do	family	go	however	lift
doctor	far	god	huge	light
dog	fast	gold	human	like
door	father	good	hundred	line
doorway	fear	grab	hurry	lip
down	feed	grandfather	hurt	listen
dozen	feel	grandmother	husband	little
drag	few	grass	I	local
draw	field	gray	ice	lock
dream	fight	great	idea	long
dress	figure	green	if	look
drink	fill	ground	ignore	lose
drive	final	group	image	lot
driver	finally	grow	imagine	loud
drop	find	guard	immediately	love
dry	fine	guess	important	low
during	finger	gun	in	lucky
dust	finish	guy	information	lunch
each	fire	hair	inside	machine
ear	first	half	instead	main
early	fish	hall	interest	make
earth	fit	hallway	into	man
easily	five	hand	it	manage
east	fix	hang	itself	many
easy	flash	happen	jacket	map
eat	flat	happy	job	mark
edge	flight	hard	join	marriage
effort	floor	hardly	joke	marry
egg	flower	hate	jump	matter
eight	fly	have	just	may
either	follow	he	keep	maybe
else	food	head	key	me
empty	foot	hear	kick	mean
end	for	heart	kid	meet
engine	force	heat	kill	member
enjoy	forehead	heavy	kind	memory
enough	forest	hell	kiss	mention
enter	forever	hello	kitchen	message
entire	forget	help	knee	metal
especially	form	her	knife	middle
even	forward	here	knock	might
event	four	herself	know	mind
ever	free	hey	lady	mine
every	fresh	hi	land	minute
everybody	friend	hide	language	mirror
everyone	from	high	large	miss
everything	front	hill	last	moment
everywhere	full	him	later	money
exactly	fun	himself	laugh	month

moon	our	quickly	send	smile
more	out	quiet	sense	smoke
morning	outside	quietly	serious	snap
most	over	quite	seriously	snow
mostly	own	radio	serve	so
mother	page	rain	service	soft
mountain	pain	raise	set	softly
mouth	paint	rather	settle	soldier
move	pair	reach	seven	somebody
movie	pale	read	several	somehow
much	palm	ready	sex	someone
music	pants	real	shadow	something
must	paper	realize	shake	sometimes
my	parent	really	shape	somewhere
myself	part	reason	share	son
name	party	receive	sharp	song
narrow	pass	recognize	she	soon
near	past	red	sheet	sorry
nearly	path	refuse	ship	sort
neck	pause	remain	shirt	soul
need	pay	remember	shoe	sound
neighbor	people	remind	shoot	south
never	perfect	remove	shop	space
new	perhaps	repeat	short	speak
news	personal	reply	should	special
next	phone	rest	shoulder	spend
nice	photo	return	shout	spin
night	pick	reveal	shove	spirit
no	picture	rich	show	spot
nobody	piece	ride	shower	spread
nod	pile	right	shrug	spring
noise	pink	ring	shut	stage
none	place	rise	sick	stair
nor	plan	river	side	stand
normal	plastic	road	sigh	star
north	plate	rock	sight	stare
nose	play	roll	sign	start
not	please	roof	silence	state
note	pocket	room	silent	station
nothing	point	round	silver	stay
notice	police	row	simple	steal
now	pool	rub	simply	step
number	poor	run	since	stick
nurse	pop	rush	sing	still
of	porch	sad	single	stomach
off	position	safe	sir	stone
offer	possible	same	sister	stop
office	pour	sand	sit	store
officer	power	save	situation	storm
often	prepare	say	six	story
oh	press	scared	size	straight
okay	pretend	scene	skin	strange
old	pretty	school	sky	street
on	probably	scream	slam	stretch
once	problem	screen	sleep	strike
one	promise	sea	slide	strong
only	prove	search	slightly	student
onto	pull	seat	slip	study
open	push	second	slow	stuff
or	put	see	slowly	stupid
order	question	seem	small	such
other	quick	sell	smell	suddenly

suggest	thick	tree	wash	window
suit	thin	trip	watch	wine
summer	thing	trouble	water	wing
sun	think	truck	wave	winter
suppose	third	true	way	wipe
sure	thirty	trust	we	wish
surface	this	truth	wear	with
surprise	those	try	wedding	within
sweet	though	turn	week	without
swing	three	twenty	weight	woman
system	throat	twice	well	wonder
table	through	two	west	wood
take	throw	uncle	wet	wooden
talk	tie	under	what	word
tall	time	understand	whatever	work
tea	tiny	unless	wheel	world
teach	tire	until	when	worry
teacher	to	up	where	would
team	today	upon	whether	wrap
tear	together	use	which	write
television	tomorrow	usual	while	wrong
tell	tone	usually	whisper	yard
ten	tongue	very	white	yeah
terrible	tonight	view	who	year
than	too	village	whole	yell
thank	tooth	visit	whom	yellow
that	top	voice	whose	yes
the	toss	wait	why	yet
their	touch	wake	wide	you
them	toward	walk	wife	young
themselves	town	wall	wild	your
then	track	want	will	yourself
there	train	war	win	
these	travel	warm	wind	

Notes

In this set, I count different word forms—like "talk," "talking," and "talked"—as one word. I also allowed most "thing" forms of "doing" words, like "talker"—especially if, like "goer," it wasn't a real word but it sounded funny. In some places, I didn't use words even when they were allowed. I could have said "ship," but I stuck to "boat" because "space boat" makes me laugh. Also, there's a pair of four-letter words that are very common, but which I left off this page since some people don't like to see them. (I didn't want to use those words anyway.)

HELPERS

A lot of people helped me with this book. Their names aren't words that people use a lot, but I'm going to write them anyway because they're important.

PEOPLE WHO KNOW A LOT OF THINGS AND TOLD ME SOME OF THEM:

Asma Al-Rawi • Edward Brash • Col. Chris Hadfield • Evan Hadfield
Charlie Hohn • Adrienne Jung • Alice Kaanta • Emily Lakdawalla
Reuven Lazarus • Ada Munroe • Phil Plait • Derek Radtke • schwal
Meris Shuwarger • Ben Small • StackOverflow • Anthony Stefano
Kevin Underhill • Alex Wellerstein • Paul R. Woche, Lt. Col. USAF (Ret.)

PEOPLE WHO HELPED A LOT:

Christina Gleason • Seth Fishman and the Gernert team,
including Rebecca Gardner, Will Roberts, and Andy Kifer
Bruce Nichols, Alex Littlefield, and the rest of the folks at HMH, including
Emily Andrukaitis, Naomi Gibbs, Stephanie Kim, Beth Burleigh Fuller, Hannah Harlow,
Jill Lazer, Becky Saikia-Wilson, Brian Moore, Phyllis DeBlanche, and Loma Huh
Richard Munroe • Glen, Finn, Stereo, James, Alyssa, Ryan, Nick,
and my helpful friends on #jumps and #computergame
And, most of all, Strong Pretty Ring-Wearer

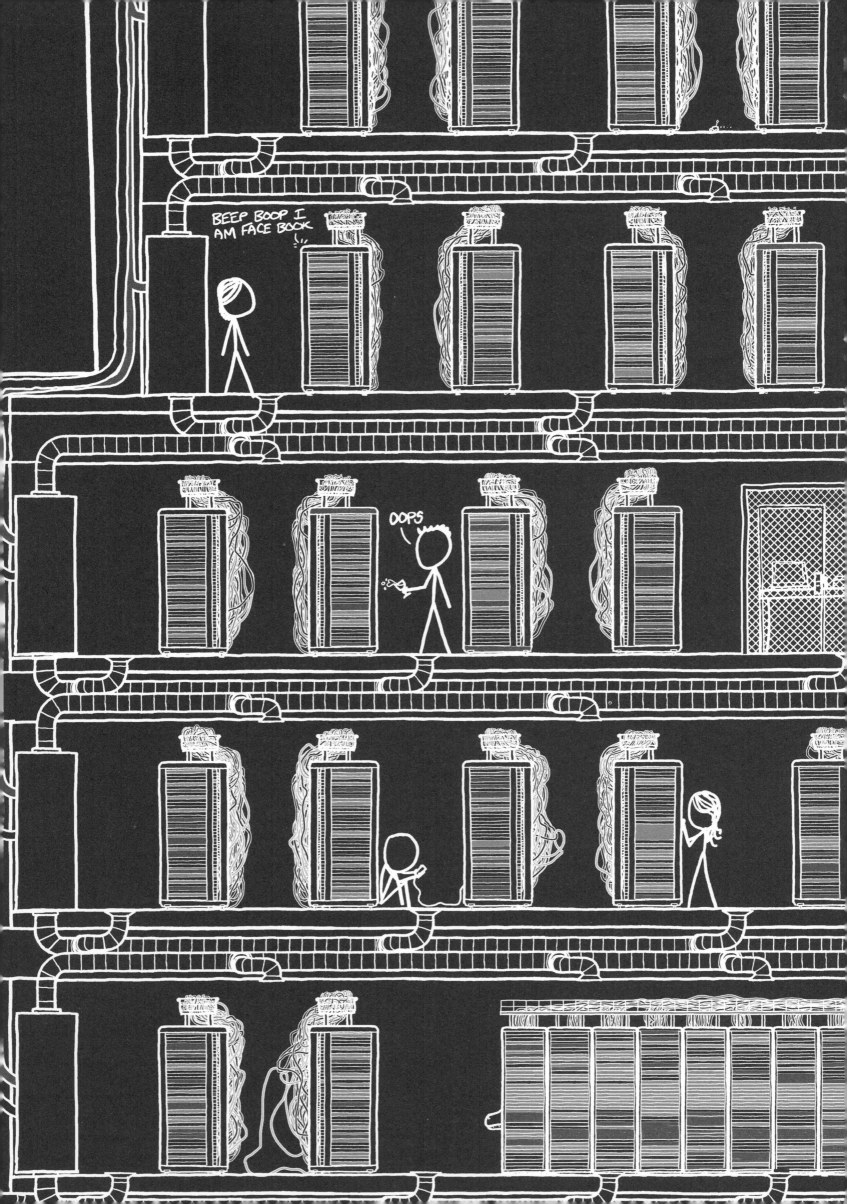